**Cost Containment and Efficiency
in National Health Systems**

*Edited by
John Rapoport, Philip Jacobs,
and Egon Jonsson*

Titles of the Series
"Health Care and Disease Management"

M. Lu, E. Jonsson (Eds.)

Financing Health Care

New Ideas for a Changing Society

2008
ISBN: 978-3-527-32027-1

S. Rashiq, D. Schopflocher, P. Taenzer, E. Jonsson (Eds.)

Chronic Pain

A Health Policy Perspective

2008
ISBN: 978-3-527-32382-1

Related Titles

S. Berger

Fundamentals of Health Care Financial Management

A Practical Guide to Fiscal Issues and Activities

2008
ISBN: 978-0-7879-9750-2

P. Muennig

Cost-Effectiveness Analysis in Health

A Practical Approach

2008
ISBN: 978-0-7879-9556-0

S. Morris, N. Devlin, D. Parkin

Economic Analysis in Health Care

2007
ISBN: 978-0-470-01685-5

Cost Containment and Efficiency in National Health Systems

A Global Comparison

Edited by
John Rapoport, Philip Jacobs,
and Egon Jonsson

WILEY-BLACKWELL

WILEY-VCH Verlag GmbH & Co. KGaA

The Editors

Prof. John Rapoport
Mount Holyoke College
Department of Economics
50 College Street
South Hadley, MA 01075
USA

Prof. Philip Jacobs
Institute of Health Economics
10405 Jasper Ave
Edmonton, Alberta T5J 3N4
Canada

Prof. Egon Jonsson
University of Alberta
University of Calgary
Department of Public
Health Science
Institute of Health Economics
10405 Jasper Ave
Edmonton, Alberta T5J 3N4
Canada

Series Editor

Prof. Egon Jonsson
University of Alberta
University of Calgary
Department of Public
Health Science
Institute of Health Economics
10405 Jasper Ave
Edmonton, Alberta T5J 3N4
Canada

Library of Congress Card No.: applied for

British Library Cataloguing-in-Publication Data
A catalogue record for this book is available from the British Library.

Bibliographic information published by the Deutsche Nationalbibliothek
The Deutsche Nationalbibliothek lists this publication in the Deutsche Nationalbibliografie; detailed bibliographic data are available on the Internet at http://dnb.d-nb.de.

Printed in the Federal Republic of Germany
Printed on acid-free paper

Typesetting Thomson Digital, Noida, India
Printing Strauss GmbH, Mörlenbach
Bookbinding Litges & Dopf GmbH, Heppenheim

ISBN: 978-3-527-32110-0

Contents

Cost Containment and Efficiency in National Health Systems: A Global Comparison
Edited by John Rapoport, Philip Jacobs, and Egon Jonsson
Copyright © 2009 WILEY-VCH Verlag GmbH & Co. KGaA, Weinheim
ISBN: 978-3-527-32110-0

Preface

The increasing cost of healthcare is of great concern in many jurisdictions. There are several measures that can be used to deal with escalating costs while maintaining or improving the efficiency of the system: these range from limits on demand and supply of health services to improved effectiveness and tighter administrative control. The variety of measures available has been tried with varying success in different settings.

Because of this profusion of economic strategies and their aftermaths, it would be helpful for policy-makers everywhere to become familiar with economic health – reform experiences and their outcomes. In this vein, the IHE sought to develop a book of comparative economic aspects of health care reform, presented in a systematic way, across a variety of countries. The book would be written by leading experts in their countries, and would provide in a readable form descriptions and analyses of recent reforms. This is the kernel that led to the present book.

Several questions guided our inquiry. Is there a convergence of economic policies among countries? Are any of the strategies that have been highlighted by the authors becoming permanent fixtures on the health economic policy scene? The absence of evidence in this area has generated questions which seem to be on the minds of policy – makers everywhere. There is a need for international scientific leadership in the economic policy arena, as there is in many clinical fields.

We could not cover every country so we chose to focus on a limited number based on the fact that each has its own national health care system, is at a similar level of economic development, and has been known to employ a variety of strategies.

We hope that this volume helps to fill in the gap of knowledge in this area, and that the systematic approach that has been taken provides a useful tool for those interested in health care reform to get a better grasp of how policy makers have attacked this complex problem.

Edmonton, August 2008

John Rapoport
Egon Jonsson
Philip Jacobs

Cost Containment and Efficiency in National Health Systems: A Global Comparison
Edited by John Rapoport, Philip Jacobs, and Egon Jonsson
Copyright © 2009 WILEY-VCH Verlag GmbH & Co. KGaA, Weinheim
ISBN: 978-3-527-32110-0

List of Contributors

Toni Ashton
University of Auckland
Center for Health Services Research
and Policy
Private Bag 92019
Auckland
New Zealand

Werner Brouwer
Erasmus University Rotterdam
Department of Health Policy and
Management
Institute for Medical Technology
Assessment
P.O. Box 1738
3000 DR Rotterdam
The Netherlands

Reinhard Busse
Berlin University of Technology
Department of Health Care
Management
Strasse des 17. Juni 145
10623 Berlin
Germany

Raisa Deber
University of Toronto
Health Policy
Management and Evaluation
Faculty of Medicine
155 College Street
Toronto, Ontario M5T 3M6
Canada

Unto Häkkinen
University of Kuopio
Department of Health Policy
and Management
Health and Welfare Economics
Siltasaarenkatu 18
P.O. Box 220
00531 Helsinki
Finland

Akinori Hisashige
Institute of Healthcare Technology
Assessment
2-24-10 Shomachi
Tokushima 770–0044
Japan

Cost Containment and Efficiency in National Health Systems: A Global Comparison
Edited by John Rapoport, Philip Jacobs, and Egon Jonsson
Copyright © 2009 WILEY-VCH Verlag GmbH & Co. KGaA, Weinheim
ISBN: 978-3-527-32110-0

Philip Jacobs
Institute of Health Economics
10405 Jasper Avenue
Edmonton, Alberta T5J 3N4
Canada

Egon Jonsson
University of Alberta
Department of Public Health Science
Institute of Health Economics
10405 Jasper Avenue
Edmonton, Alberta T5J 3N4
Canada

Bengt Jönsson
Stockholm School of Economics
Handelshögskolan i Stockholm
Box 6501
Sveavägen 65
113 83 Stockholm
Sweden

Adam Oliver
The London School of Economics
and Political Science
Houghton Street
London WC2A 2AE
United Kingdom

John Rapoport
Mount Holyoke College
Department of Economics
50 College Street
South Hadley, MA 01075
USA

Frans Rutten
Erasmus MC (University Medical
Center Rotterdam)
Institute of Health Policy and
Management (iBMG)
Institute for Medical Technology
Assessment (iMTA)
P.O. Box 1738
3000 DR Rotterdam
The Netherlands

Markus Wörz
Berlin University of Technology
Department of Health Care
Management
Strasse des 17. Juni 145
10623 Berlin
Germany

1
Introduction and Summary

John Rapoport, Egon Jonsson, and Philip Jacobs

1.1
Introduction

The chapters in this book review several decades of developments in health policy related to cost containment and efficiency in eight different countries. Both the similarities and the differences among the countries are striking. Table 1.1 documents their population, health spending and health status using OECD (Organisation for Economic Co-operation and Development) data. While these countries are not a random sample of OECD countries, it is notable that for all the characteristics included in the table the eight countries come from both halves of the OECD group of 30 countries. That is, they are found both above and below the OECD median values.

All the countries included in the book exist in the same global economy, have access to essentially the same medical technology and information, and are roughly at the same stage of economic development. The fundamental economic problems and sources of potential market failure affect health systems in all of them. However, the policy responses differ because of historical and cultural differences, varying political ideologies and social values, differences in paths of evolution of the national health systems, and the health needs of the population. In this chapter, we attempt to summarize the strategies discussed in the book, their effects on cost containment and efficiency, and their success and sustainability.

In the first section we describe briefly the approaches presented by each chapter author, and then discuss the conceptual relationship between cost containment and economic efficiency. We then examine the details and success of each strategy in a cross-country analysis. The concluding section considers the current policy agenda.

1.2
Highlights of Each Country's Approach

Canadian health systems are organized at the provincial level, but operate within a set of national principles and obtain significant funding from the national government.

Cost Containment and Efficiency in National Health Systems: A Global Comparison
Edited by John Rapoport, Philip Jacobs, and Egon Jonsson
Copyright © 2009 WILEY-VCH Verlag GmbH & Co. KGaA, Weinheim
ISBN: 978-3-527-32110-0

Table 1.1 Statistical profiles of countries included (data is from 2005).

	Canada	Finland	Japan	Sweden	UK	Germany	Holland	New Zealand	OECD median
Population (million)	32.3	5.2	127.8	9.03	60	82.5	16.3	4.1	10.8
GDP per capita (US$, PPP)	33 600	33 300	30 500	32 700	32 100	29 800	34 200	25 300	31 000
Health spending as % GDP	9.8	7.5	8	9.1	8.3	10.7	9.2	9	9.05
Public health spending as % of total spending	70.3	77.8	81.7	84.6	87.1	76.9	62.5	78.1	75
Health spending per capita (US$, PPP)	3326	2331	2358	2918	2724	3287	3094	2343	2922
Acute care beds per 1000 population	2.9	2.9	8.2	2.2	3.1	6.4	3.1	NA	3.4
Practicing physicians per 1000 population	2.2	2.4	2	3.4	2.4	3.4	3.7	2.2	3.4
Life expectancy at birth (years)	79.9	78.9	82.1	80.6	79	79	79.4	79.6	79.3
Infant mortality per 1000 live births	5.3	3	2.8	2.4	5.1	3.9	4.9	5.1	4.1

NA = not available.
Sources: OECD Health Data, 2007; OECD in Figures, 2007.

While there is significant variation among provinces, Raisa Deber identifies several common strategies for both cost containment and efficiency enhancement. Supply-side measures include capped budgets for hospitals and physicians, as well as policies to limit the supply of physicians and other health care workers. Demand side policies include measures to address the appropriateness of care and, to a limited extent, to promote health and prevent disease. Organizational and structural initiatives to provide integrated care and to encourage competition are also discussed.

In his chapter on England, Adam Oliver makes clear that efficiency seeking has been the major policy goal over the past 20 years. This concern has led to changes in both the structure and operations of the National Health Service. The introduction of an internal market in 1991 enabled purchasers to negotiate contracts with competing providers of health care. The establishment of the National Institute for Health and Clinical Excellence (NICE) during the late 1990s provided analytical expertise to make information provided through health technology assessment available to decision makers, and positive NICE guidance was made mandatory for the National Health Service (NHS) in 2001. More recent reforms reviewed in the chapter include performance management techniques focused on reducing waiting times and increased opportunity for patients to choose their hospital.

In the chapter on Finland, Unto Häkkinen notes that cost containment first became a major concern as a result of the economic downturn of the early 1990s. The policies to address this worked through the decentralized nature of the Finnish system as the central government reduced its specific control and municipalities took more responsibility for organizing services. Government information guidance to municipalities through health technology assessment, improved statistical systems and strategic planning played an important role. In the pharmaceutical sector, the regulation of wholesale drug prices and generic substitution are discussed.

Cost containment has been the major objective of policy in Germany for the past 30 years, with efficiency also being an important focus since the 1990s. Markus Wörz and Reinhard Busse detail the role that global budgets and spending caps have played in the ambulatory care, hospital and pharmaceutical sectors. They explain the trend towards case-based hospital reimbursement. Cost shifts to private households through cost sharing and benefit exclusion are also discussed. The introduction of competition between sickness funds and providers led to major structural changes which are examined. Finally, the chapter explains the increased importance of joint self government in the German system.

The ongoing shift from supply-side policies to demand limitation is presented by Werner Brouwer and Frans Rutten as the context for their discussion of health system reforms in Holland. Supply-side policies in effect during the 1980s included price regulation, budgets and waiting lists. In the pharmaceutical area, reference pricing in various forms has been the major policy with measures to affect prescribing behavior and value-based reimbursement playing a greater role as the system evolves. In the post-2000 period, policy has been oriented towards encouraging competition, and specific strategies to limit demand include cost sharing and limits on the basic benefits package.

In his chapter on Japan, Akinori Hisashige makes clear that cost containment has been the focus, with efficiency being a much less prominent policy objective. Budget setting at the national level, combined with a national fee schedule for payments to all providers and control on drug prices, has been the main approach to achieving the cost-containment objective. Fixed-bundle payments for sets of services have also been used. Controls on hospital beds and length of stay, health care manpower policy, and patient cost-sharing have been attempted but are of much less importance than national budget setting.

New Zealand's publicly funded health system first became concerned with efficiency during the 1980s. In her chapter, Toni Ashton discusses five policy strategies which together address both cost containment and economic efficiency concerns. A national global budget for health as well as regional budgets has been part of the system since early in its history, but recent changes in the way that accounting is carried out and budgets are enforced has made them a more important tool during the past 15 years. The quasi-market reforms of the 1990s described by Ashton involved a purchaser–provider split and significant restructuring of the organizations involved in the funding and provision of care. A national agency to manage pharmaceuticals was established. Waiting list management was also the focus of a specific policy. Priority settings for spending, using techniques such as clinical guidelines and technology assessment, were implemented.

Sweden has a very decentralized health care system with most operational decisions made at the regional or county council level. Bengt Jönsson points out that, as the system has evolved from pluralistic to single payer, the decentralization has been maintained while cost containment and efficiency have both played a role as major goals of policy. The diagnosis-related group (DRG)-based performance payment of hospitals, which was first discussed during the 1980s and more fully implemented in the 1990s, has implications for both goals. Sweden was one of the first European countries to establish a formal institute for Health Technology Assessment (HTA), and Jonsson traces the role that HTA has played over the past 20 years. Pharmaceutical policies are also a main focus of discussion, as Sweden has had experience with both reference pricing and generic substitution of drugs.

1.3
How Are Cost Containment and Efficiency Related?

The term 'cost containment' as used in most discussions in this book refers to reducing or slowing the rate of growth of health care spending. Sometimes, the reference is to health care spending by the government, while in other cases the concern is with overall national health care spending, whether government or private sector. 'Efficiency' implies spending money on that set of uses which yields maximum benefits. It is important to bear in mind that the reason the level of health care spending matters for social welfare is that the money spent on health care has alternative uses. It is *whatever else* could have been produced with this money that represents the true cost (i.e. the opportunity cost) of health care. This link between spending and opportunity cost is why it is important to consider economic efficiency alongside cost containment. Policies directed at cost containment may have intended or unintended effects on efficiency; policies seeking efficiency may raise or lower health care costs.

Lavis and Stoddard present a useful economic framework for thinking about efficiency [1]. As they note, there are "... three levels of efficiency in economic theory: technical efficiency, cost effectiveness and allocative efficiency. Production is technically inefficient or cost-ineffective if the same 'output' could be produced with, respectively, fewer or less expensive 'inputs'; production is allocatively inefficient if an equally valued level or mix of output is possible using fewer resources" ([1], p. 46). In summary, efficiency is both about 'doing things right' and 'doing the right things'.

Economic efficiency may not be the primary goal of government decision makers. Distributional effects are implicit in any policy change. Sometimes, a distributional issue is a stated goal, such as in New Zealand where Maori health is an explicit focus of government policy, or in Sweden where according to health care law 'human values' and 'equity' must be considered together with cost-effectiveness as the guiding principle for resource allocation. More commonly, differential effects on groups in society are simply part of the political process leading to policy change. Typically, any policy change is likely to have both efficiency and distributional effects. From an analytical standpoint, the key aspect of an efficiency change is a difference in economic behavior. If one can identify specific incentives, which economic actors they affect, and how those actors change behavior resulting in different resource allocation, then one is talking about efficiency.

1.4
Strategies and Their Effects: A Cross-Country Analysis

The list below shows, in summary form, the strategies for cost containment or improved efficiency which the chapter authors have identified. It should be noted that this does not represent a comprehensive or complete list of all possible approaches. There may be other policies which were used in minor areas of health systems or which did not change during the time period considered. It should be noted that only those strategies which the authors brought forward as most important in their country's policy over the past few decades for addressing the issues of cost containment and efficiency are included in the list:

- **Related to information for decision-making**
  - Analytical process (such as HTA or cost-effectiveness analysis) to guide decisions about the services included
  - Strategic planning

- **Related to price regulation**
  - DRG-type hospital reimbursement
  - Regulation of physician fees
  - Reference pricing of drugs

- **Related to budget setting or supply limitation**
  - Global budgets (entire sector or large part)
  - Capped budgets (specific providers or services)
  - Limit human resource supply

- **Related to financial incentives for individuals**
  - Cost sharing with households for covered services

- **Related to creating market incentives**
  - Increased opportunity for individual to choose insurer or hospital
  - Purchaser-provider split/internal market

- **Related to specific aspects of delivery system**
  - Increased size of insurers/regionalization
  - Waiting list management
  - Performance management
  - Generic substitution of drugs

- **Other**
  - Health promotion/disease prevention

Good *decision-making* requires good information. It is widely recognized by economists that information problems can lead to market failure and that provision or regulation of information is an important government function [2]. So, it is not surprising that government activities in the area of research and analysis, such as economic evaluation, HTA and health policy research figured prominently in the chapters for many countries. Indeed, seven of the eight countries identified government agencies or programs in this area. Most focused on new technologies and the question of which services should be included in the benefit package. A number of authors mentioned the importance of such a role for government since private interests are very active in lobbying for influence and producing analysis to support such lobbying. The main differences among countries are the extent to which a formal analytical framework, such as cost–utility analysis or cost–effectiveness analysis, was relied upon, the degree of independence of the agency from the main health system leadership, and the extent to which recommendations of the agency were mandatory or simply advisory. At one extreme perhaps is England's National Institute for Health and Clinical Excellence (NICE), which relies extensively on cost–utility analysis and produces guidance which, in the case of a positive recommendation that an intervention is cost-effective, is mandatory for providers to follow. Although Sweden was one of the earliest countries to establish a HTA agency, it does not make recommendations which are mandatory for system participants to follow. In Finland, in recent years, increased funding has been provided to the FinOHTA, but its role remains advisory. In Germany, the information function is located in an institute integrated with joint self-governance, that is a body with provider, insurer and (nonvoting) patient representation. Holland is at a fairly early stage of thinking about the possible role of cost–utility analysis in deciding what to include in the benefit package. Canada has established a national agency for HTA with an advisory role, while Finland and New Zealand each have broader priority setting or strategic planning functions which touch on the issues of technology assessment embedded in a more general framework.

Little clear evidence was provided on the success of information-related strategies. In Holland, the use of cost-effectiveness analysis for benefit package definition and in New Zealand a new framework for HTA were too recent to have been evaluated. The authors of the chapters on Canada, Finland and Sweden each noted that HTA seemed to have little effect on the actual health system operations because there was no mandate that information be considered or recommendations followed by decision makers. As Raisa Deber put it, the HTA agencies had "... very few policy levers to translate their recommendations into policy". Bengt Jonsson attributed the absence of effect of HTA on policy to "... lack of a systematic approach in the health care system to introduce and evaluate new technology". In Germany, Worz and Busse note that economic evaluation tools are used in decisions about individual technologies. By other criteria, HTA in some countries was very successful. Both Adam Oliver in England and Bengt Jonsson in Sweden pointed to the high scientific quality of studies by NICE and the Swedish Council on Technology Assessment in Health Care (the SBU), respectively, and the international recognition given to those agencies. In England, of course, the guidance of NICE is in part mandatory, so it did have some effect on system operations. However, Oliver details a number of methodological issues and criticisms which continue to generate controversy. One of particular interest in light of the twin foci of this book is the finding that a new technology can not only be cost-effective but also cost-increasing. In this case, a policy decision about whether to adopt the technology implies a choice between the policy goals of cost containment and economic efficiency.

The use of case-based, DRG-type, hospital pricing was one of the most mentioned strategies, and was used by six of the eight countries. In most countries it was initiated at or after the mid 1990s. This type of hospital reimbursement provides an incentive for hospital managers to treat patients at the lowest possible cost and also to treat more patients, as long as a hospital objective is to increase its surplus or reduce its deficits. While this certainly is consistent with reducing technical inefficiency, the cost reduction could also be achieved by diminishing quality; for example, reducing the length of stay below the clinically optimal level. The primary objective of this reform was somewhat different in each case. In several chapters the authors noted the key role of the switch to DRG-type hospital reimbursement, when combined with patient choice, in fostering competition among hospitals. Two types of competition were identified, each with an implicit technical efficiency result. In England, and to some extent in Sweden, competition among hospitals on the basis of quality might be facilitated since, with every hospital receiving the same case-based price, raising quality would represent one way to attract patients. In Holland and Germany the emphasis was more on the ability of case-based pricing to enable competitive negotiation over prices between insurers and hospitals. In Finland, the fact that DRG prices would better reflect real costs was seen as helpful in assuring equity among municipalities in hospital funding. Japan's use of DRG pricing was a pilot program aimed at cost containment as the primary goal.

Evidence on the success of case-based hospital pricing varied considerably among countries. In Sweden, Bengt Jonsson reported that no change in quality was apparent and a clear improvement in hospital productivity was found. Also in Gemany there was evidence of increased hospital efficiency when case-based pricing was introduced. Akinori Hasashige reported from Japan that results from the small-scale use of DRG-type pricing were not promising for cost containment. In England, Germany and Holland it is too soon to make an evaluation of choice of hospital as facilitated by case-based pricing.

Price regulation was used for services other than hospitals as well. In Japan, a revision of the comprehensive fee schedule, covering all services, was the primary mechanism of cost containment. In the pharmaceutical area, Sweden, Holland and New Zealand have each used reference pricing for drugs in slightly different forms and with different results. A *reference pricing system* sets the price that an insurer will pay for a drug at the level of the lowest drug in the therapeutic class (or, in the case of Holland, at the average price of drugs in the therapeutic cluster). For a drug priced above that level, the patient will have to pay the difference between that price and the reimbursable price. This of course creates a strong incentive for patients not to choose higher-priced drugs and for drug manufacturers to price at no higher than the reimbursable level. Potential problems include administrative complexity, a decreased incentive for the introduction of new drugs in the country, and no incentive for competition to drive the price below the reimbursable price. Sweden abandoned the reference pricing system in 2002 in favor of a policy of generic substitution because reference pricing was administratively complex, did not achieve long-term savings, and also reduced the entry of new drug sellers. Holland excludes some drugs from the therapeutic cluster. The system seems to work well in New Zealand for cost containment, although there is some concern there about access to new drugs. Finland directly regulates the price of prescribed medicines, with wholesale prices determined administratively.

Supply limitation, which can include global or capped budgets as well as limits on particular resources, is a very direct way to address the cost-containment objective. Budgets have played an important role in the policy of many countries. Typically, countries vary in the extent to which the budget applies to the whole system or to a specific type of provider, how hard or fixed the budget cap is, and how the budget level is determined. In Japan, a budget ceiling set by the Ministry of Finance for the health ministry is a key point of control. There is also a national budget in New Zealand which covers about 65% of total health expenditures, and is allocated to district health boards on a population basis. Currently it is a fairly hard limit, with financial penalties for deficits.

Both Canada and Germany, and also Holland until fairly recently, used capped budgets for individual hospitals as cost-containment strategies. In Canada this led to significant decreases in numbers of hospital beds during the 1990s, in attempts by hospitals to shift costs to nonhospital providers, and also in waiting lists for hospital services. In Germany, capped budgets combined with progress toward case-based pricing for hospital services were associated with some attempts by hospitals to shift costs, to a decreased length of stay and to greater technical efficiency. A cap on total reimbursement to physicians for ambulatory care was also used in Germany, and also for a short time in most Canadian provinces, combined with a fee for service compensation of individual doctors. The efficiency concern here is similar to a fee for service without a capped budget – incentives for physicians to provide more service. With a binding cap, however, the issue is not the concern of 'too much' care provided but rather the combination of particular services. Both countries thus introduced payments systems which addressed differences in compensation for varying bundles of services or types of care. Spending caps for drugs in Germany and in New Zealand during the 1990s played a significant role, and were fairly effective methods of cost containment in that area. In Finland, global budgets at the local level have effectively controlled health expenditures with elected local government responsible for health as well as other public services and relying significantly on local taxes.

The efficiency implications of a global or capped budget approach depend on how the budget is set, how strictly the budget limit is enforced, and what arrangements are in place to distribute the budget among providers. There may be incentive effects growing out of methods to set the level of the budget. For example, if next year's budget allocation depends on the current year spending there may be greater likelihood of increased spending this year. But this could in fact reward inefficiency. In addition, if it is possible to shift costs or functions from one governmental agency to another, or to the private sector, it may be possible to escape a budget constraint. Such budget shifting was experienced at times in New Zealand, Canada and Japan.

In general, budgets were reported to be effective ways to contain costs. For example, Wörz and Busse report that pharmaceutical spending caps in the 1990 "... proved to be an effective method for a short-term reduction and long-term modification of pharmaceutical expenditure", and that spending as a percent of GDP has remained stable for hospitals and physician practices. In New Zealand, Toni Ashton notes that "... global budgets for publicly funded services have been important historically in containing total health expenditure", and "... have become increasingly effective in containing costs in recent years".

Specific controls on the supply of doctors and nurses were also reported. Canada, in reacting to projections of oversupply, imposed tighter restrictions on the ability of foreign medical graduates to obtain licenses to practice in Canada, and reduced the number places available in medical schools. Nursing school enrollment was similarly reduced. Japan's manpower policy, on the other hand, has attempted to increase the number of physicians, while constraining growth in the number of hospital beds.

Attempts to inject elements of *market competition* into the health care system were made in many countries during the 1990–2008 time period. Typically, these involved a separation of the role of purchaser and provider, an increase in the degree of choice for individuals or organizations and, in some cases, an enhanced openness to private market participation in what formerly had been an almost totally government activity. In England, the NHS was restructured, with District Health Authorities and general practitioner (GP) fundholders (later Primary Care Trusts; PCTs) becoming purchasers of hospital and other health services from providers, but now including those in the private sector. New Zealand centralized the purchasing function into a single Health

Funding Authority which would purchase hospital services from not-for-profit government owned providers. Sweden and Finland, each with very decentralized systems, freed the governmental subunits (county councils and municipalities respectively) to become purchasers and contract for hospital and other services. Germany and Holland increased choice in the system at the individual–insurer interface, the former by freeing individuals who had previously been assigned to a sickness fund based on occupation or location to choose which sickness fund to join, and the latter by introducing national health insurance operated by competing private insurers and freedom of switching for every insured at the turn of the year.

Market power, due to either buyers or sellers being large relative to the total market size, was a problem mentioned by the authors of several chapters as a reason why market reforms might not in fact attain the desired results. In England, there was concern that the large size of hospitals would limit the ability of much smaller GP fundholders to bargain successfully for lower prices. In Sweden, the small number of hospitals in local market areas also limited the ability of the market to reduce prices. If low availability of hospitals in a local market and travel costs make it unlikely that a patient can realistically choose among several hospitals, the theoretical possibility of choice is not meaningful. On the other hand, market power on the buyer's side – which tends to depress price – can be a favorable factor from a cost-containment standpoint in a market system.

Proposals or policies for market solutions ran into institutional or political barriers. In countries with a strong emphasis on equity and a long tradition of public funding and government provision of health services (e.g. Sweden), market-oriented reforms could been seen as conflicting with the basic value system. In other cases, established relationships and patterns of behavior such as the role of the municipality as both potential 'shopper' for hospital services in a market but also the operator of its own local hospital in Finland, or the collegial referral networks developed under the NHS in England, made it difficult for providers and health system officials to change their ways of thinking to that of buyers and sellers in a market.

Market-oriented reforms were seen by most authors as of only limited success. Toni Ashton writes from New Zealand that ". . . the quasi-market structure that was in place in New Zealand from 1993–1999 was less effective in achieving efficiency gains than its proponents expected". A similar conclusion from England by Adam Oliver noted that ". . . a case can be made that the internal market reform of the 1990s had only a short-term effect on productive efficiency." Evidence from Germany is less clear. Wörz and Busse document that substantial changes in the structure of the sickness fund market have occurred as well as changes in contributions and the use of integrated care contracts. However, the ultimate effect on costs, or on health outcomes, awaits further evidence. Market-oriented reform seems to be successful, at least temporarily, in Holland with a leveling off of premium increases.

Although many countries had some fees for specific services (e.g. in Sweden, where copayments are important for drugs and dental care), changes in direct payments by households played a relatively small role in health care reform in the countries reviewed in this book. Four countries experienced developments in this area. In both Germany and Japan, household payments for insurance coverage as well as copayments for services were increased during the period between 1980 and 2000. In Finland, household payments were increased during time of economic recession. Holland experimented for several years (2005–2008) with an innovative 'no-claim rebate', that is a refund to individuals who did not use a large amount of health services, but abandoned the system as incentives to consumers seemed to be weak.

Policies aimed at sharing the cost of health care between the government and individuals and placing a higher burden of financing on individuals can have several different types of effect. To the extent that a larger share of costs are paid by individuals and less by the government, it clearly advances the cost-containment objective (assuming that 'cost containment' is interpreted as reducing government spending on health care). Fees for individuals at the point of receiving care can also reduce demand for care. There are also significant distributional effects from cost sharing. Copayments obviously affect those who seek care rather than those who do not, and so they can be seen as a shift of the economic burden from those who are well to those who are sick. A given copayment may also be more of a burden for a low-income person than for a person with a higher income. Both of these effects may be considered inequitable and incompatible with solidarity. The administrative costs of cost-sharing policies also tend to be high.

Experiences with cost sharing were rather different in the countries which tried policies of this sort. In Holland, it was not possible to implement policies which were seen as both equitable and effective in reducing medical care utilization, so these strategies were abandoned. In Germany, cost sharing was increased during the 1990s through 2003, although research on their economic effects of this proved to be inconclusive. In Japan there were large increases in the employee share of health insurance premiums over the 1983–2003 time period and copayments for a variety of services. Akinori Hisashige noted that demand changes resulting from the higher effective price were small there, and thus the approach is widely used as a cost-containment policy.

Changes in the size of governmental, insurance or operational health care units formed part of some reforms. In Finland, recommendations made during the 1990s suggested the need for bigger municipalities to increase the size of population included in a health care coverage unit, and in 2005 the government began a specific initiative to achieve this. Many Canadian provinces moved towards the regionalization of hospital services. In Germany, although it was not a direct aim of a policy but rather a reaction by sickness funds to increased competition, the average size of sickness funds was increased. In New Zealand, regional authorities were replaced with a single national agency for purchasing health and social care services.

The reforms to increase the size of insurers or government funding units may well have been attempts to increase the buyer's ability to negotiate a favorable price. For example, in Germany Wörz and Busse mention that one reason for the increase in size (through merger) of sickness funds is a ". . . gain in bargaining power of sickness funds in negotiations with providers." Similarly, in New Zealand the main reason given by Toni Ashton for the establishment of Pharmac, the national pharmaceutical management agency which she characterized as ". . . spectacularly successful in controlling government expenditures", was ". . . to achieve economies of scale through joint purchasing." Economies of scope – the cost saving from performing several functions jointly in the same organization – can also come from larger size. Disease management programs in larger sickness funds in Germany and larger Finnish municipalities are possible examples of such effects.

The above discussion has summarized evidence on each of the various strategies highlighted in the chapters of this book. While such a review is very useful it has at least one major limitation, however, since in order to fully understand the implications of a strategy it must be considered in context. The culture and values of a country importantly affect which policies will be adopted, and whether they will be successful. Policies must be evaluated not individually but rather in combination. Participants in the health care system react to the mix of incentives in place at a given time. The effect of a specific policy may be very interdependent with other policies. An example here might be the effect of case-based DRG-type hospital payments on quality. If a sole large buyer of hospital services changes the method of payment from cost-based to case-based, the incentive for hospitals to cut costs may result in lower quality of care – that is, discharging patients 'sicker and quicker'. On the other hand, case-based payment mandated for multiple buyers of hospital services, where hospitals have to compete for business, was thought to generate quality competition and thus improve quality – an opposite result.

Also with physician payment and payment for drugs, the whole combination of policies must be considered. Fees for service payment in a relatively unconstrained system are likely to increase spending on physician services due to the possibility of supplier-induced demand [3]. The volume of primary visits would increase and referral networks might increase specialist visits. On the other hand, fees for service payments in a system with a tight budget cap on the total physician payments create a situation where doctors are rivals for shares of the fixed income 'pie'. Less referral activity and rivalry between specialties might be expected. Finally, with a drug policy the need to look beyond a specific policy instrument was evident. Reference pricing and generic substitution had different effects on cost containment in systems where the retail pharmacy markets were competitive compared to those in which these markets were monopolized. For example, the success of generic substitution in Sweden was attributed by Bengt Jonsson to the fact that retail pharmacy was a public monopoly, whereas in Holland the margin competition at retail pharmacy level was reported by Brouwer and Ruttan to lead to higher drug expenditures.

1.5
Current Policy Agenda

Most authors of chapters in this book could not foresee any major change of direction of policy in their country in the immediate future. Access to care and quality of care are high priorities, while rising costs are a continuing concern, even if specific cost-containment policies are not to the forefront. Indeed, although cost containment was not the primary objective of policy in most countries during the time period reviewed, several authors noted that this could change with a deterioration of the macroeconomic environment. Efficiency was an important goal for most countries, and in recent years most have adopted policies which appear to have made some progress in achieving increased efficiency. Concern was expressed that some of these achievements might be one-time or short-term effects, and the task of maintaining efficiency and affordability in the face of changing medical technology is seen as a major challenge. In particular, policies regarding pharmaceuticals are a current focus and concern. Most countries have recognized that whilst advancing technology has considerable potential to improve health outcomes and economic efficiency, it is also a major driver of increased costs. Policies to address this dilemma include methods for the approval and pricing of new drugs and health technology assessment.

The difficulty of targeting population health status with policy, in contrast to health care alone, was noted by several authors. Health promotion and disease prevention are ways to address this, as well as environmental improvement and lifestyle changes. However, such policies often require intersectoral approaches and cooperative efforts by agencies beyond those primarily responsible for health care. These are often particularly difficult to organize and implement.

It was recognized that political support for any national health system depends ultimately on a perception that the system is fundamentally fair. *Equity* was an underlying theme in most chapters, both in the review of past policies and in consideration of the current agenda. However, there were clear differences in the way this was conceptualized. A focus on health outcomes, and reducing differences in health outcome, was an objective of New Zealand's future efforts, while in England there was concern that the efficiency emphasis central to NICE analysis did not consider equity concerns. Holland's policies in the area of cost sharing were abandoned partly because it was seen as unfair. In Germany, the system is moving towards a uniform contribution rate for insurance, rather than have differences between sickness funds.

On balance, the trends in these countries seem consistent with the 'third wave' of health care reform identified by David Cutler [4]. That is, there was less emphasis on supply-side limits and greater focus on market-like incentives for changes in behavior, such as pricing and increased opportunities for choice Beyond that, it is difficult to discern a consensus or convergence on specific policy approaches. Some countries have abandoned policies which others are just beginning to introduced. Clearly, what is judged as successful in one country may be deemed unacceptable in another.

But what could be the reason for such an apparently confused picture? There are at least three possible (albeit not mutually exclusive) reasons:

- Countries may have different political and economic goals; some may consider cost containment and efficiency as equally high priority objectives, while others give higher weight to one or the other.
- The social and cultural setting may affect the ability to implement a policy and its chance for success.
- What we are seeing here is a dynamic learning process – a pattern of trial and error which, over time, will yield information about which approaches will achieve their

intended goals and how they may have unintended (favorable or unfavorable) results.

To the extent that such a learning process is underway, the international exchange of information and sharing of experiences is highly beneficial. A given country can learn from the results of another country's experiment. As active, well-informed discussion and debate is an essential part of this learning process, hopefully this book will contribute to the debate and challenge of containing costs and enhancing efficiency in national health systems.

References

1 Lavis, J.N. and Stoddart, G.L. (1994) Can we have too much health care? *Daedalus*, **123** (4), 43–60.

2 Pindyck, R.S. and Rubinfeld, D.L. (2005) *Microeconomics*, 6th edn, Pearson Prentice-Hall, Chapter 17.

3 Reinhardt, U. (1999) The economist's model of physician behavior. *The Journal of the American Medical Association*, **281** (5), 462–5.

4 Cutler, D.M. (2002) Equality, efficiency, and market fundamentals: the dynamics of international medical-care reform. *Journal of Economic Literature*, **40** (3), 881–906.

2
Canada

Raisa Deber

Population (million)	32.3
GDP per capita (US$, PPP)	33 600
Health spending as % of GDP	9.8
Public health spending as % of total spending	70.3
Health spending per capita (US$, PPP)	3326
Acute care beds per 1000 population	2.9
Practicing physicians per 1000 population	2.2
Life expectancy at birth (years)	79.9
Infant mortality per 1000 live births	5.3

Strategies used

1. Capped budgets
2. Health human resources limitations
3. Organizational approaches (increase integration, encourage competition)
4. Increase appropriateness of care
5. Health promotion/disease prevention

2.1
Canada's Health Care System

Today, Canadians are both proud and worried about their 'health care system'. The quotation marks around the term reflect the fact that, as noted below, one cannot really speak of a Canadian health care system. About 70% of health expenditures are paid from public sources – including almost all medically necessary physician and hospital costs. However, even within these publicly funded sectors, Canada uses what the OECD calls a *public-contract model* [1], which means that services are delivered almost entirely by private providers, thus introducing substantial variation across providers. Because constitutional responsibility for health care rests at the subnational (provincial/territorial) rather than at the national level, there is also considerable variability

Cost Containment and Efficiency in National Health Systems: A Global Comparison
Edited by John Rapoport, Philip Jacobs, and Egon Jonsson
Copyright © 2009 WILEY-VCH Verlag GmbH & Co. KGaA, Weinheim
ISBN: 978-3-527-32110-0

across jurisdictions as to how services are managed and delivered. Yet, amid the variations there are many commonalities.

By most measures, health care in Canada has been a rousing success. Although there is always room for improvement, Canadian health outcomes exceed OECD averages, and have continued to improve [2, 3]. The system is widely admired internationally; indeed, it is ranked relatively highly by those who need to receive care [4, 5]. Costs are approximately what would be expected given Canada's wealth [6], with relatively good cost control [2, 7]. Yet, since its inception there has been a constant drumbeat of worry among Canadians that the system is financially unsustainable, paradoxically coupled with the argument that yet more money needs to be spent [4, 7]. Policy responses have swung from a narrow focus on cost containment to a similarly narrow focus on improving access and shortening wait times for specified procedures. Throughout, the advocates of a larger private role have persistently – although to date unsuccessfully – argued for a massive restructuring of how care is financed and delivered [8]. Although the ultimate result is still 'up for grabs', it provides an illustration of the difficulties in translating policy prescriptions recommended by health policy analysts into sustainable policies. The Canadian case demonstrates that, to the extent that health policy often requires trade-offs among incompatible outcomes, policy makers and the public may find it difficult to accept the resulting ambiguities. Incremental improvement does not generate immediate political rewards, and complex mixed systems are difficult to explain in sound bites.

The strategies used in Canada began with efforts to limit available resources, with an admixture of efforts to increase efficiency. For the most part, both legal restrictions and public opinion have limited the use of financial incentives directed towards potential users of health care. The decentralized nature of delivery has restricted the levers available to payers and left administrative controls on service use in the hands of individual service providers; in turn, their impact is related to the incentive structures inherent in how these providers are paid. Accordingly, the key strategies employed to contain costs and/or improve efficiency, as analyzed in this chapter, included several *supply-side* measures, including:

- Capped budgets for hospitals and physicians;
- Health human resources limitations directed at physicians and other health workers, including nurses.

Governments have also employed *organizational* approaches, including efforts to both increase integration and encourage competition.
Finally, the *demand-side* measures included:

- Efforts to increase the appropriateness of care;
- Limited health-promotion/disease-prevention activities.

2.1.1
Paying for Canadian Health Care

Although, in total, approximately 70% of Canadian health expenditures come from public funds (this is among the lower shares for OECD countries), such expenditure is highly sector-specific. Approximately 99% of physician expenditures, and 90% of those of hospitals, are derived from public sources, as compared with 4% of expenditures for dentists [9]. One major complexity is that Canada does not – and cannot – have a national health care system. Constitutionally, responsibility for health care rests at the subnational level with the 10 provinces and three sparsely populated northern territories; views of federalism, and the extent to which it is appropriate for the national government to interfere in an area under provincial jurisdiction, therefore complicate the debate and restrict the scope for national programs [10–17]. Accordingly, the use of specific strategies often varied by jurisdiction, both in timing and nuance. In contrast, under Canada's constitution, regional/local governments within provinces have no independent authority, although responsibility for particular services may be downloaded (or uploaded) as the provinces see fit [18, 19].

Because there is a considerable disparity in fiscal capacity across provinces, it has long been recognized that national levels of services would require a redistribution of resources; Canada has dealt with this through *fiscal federalism* [20]. Health care financing has thus employed agreements that the provinces/territories would provide universal insurance for a basket of services, including – but not restricted to – all medically necessary physician and hospital care, and to be financed (in part) through transfer payments from the national (federal) government [21].

The evolution of these transfers has been both complex and contentious; the federal government's view of the details is available from the federal Department of Finance web site [22]. The battles involved trade-offs among predictability, accountability and autonomy [23]. The 1957 Hospital Insurance and Diagnostic Services (HIDS) Act provided federal funds to share the costs (on a roughly 50:50 basis) for hospital-based services; the 1968 Medical Care Act added cost-sharing for physician services, and by 1972 all provinces and territories had set up complying plans. However, constant debates as to the amount of money, and the strings to be attached to federal payments, led to the replacement of these cost-shared programs (plus a similar program for post-secondary education) in 1977 by the new Established Programs Funding (EPF). The new formula added predictability; federal transfers were no longer based on actual provincial expenditures, but rather on population and inflation.

The EPF represented a shift from cost-shared models where the federal transfers were entirely in cash, to a more hands-off approach. In this new approach, the federal government agreed to reduce its income and corporate tax rates, leaving the provinces with room ('tax points') to increase their own tax takes without increasing the total tax burden on their residents. Once provided, these tax points could not be taken back. The federal government retained some (albeit substantially diminished) leverage over provincial spending through the use of 'residual cash transfers', which were designed to make up the difference between the formula's entitlement and the amount yielded by provincial taxation. This has caused enduring confusion as to what should be deemed part of the federal transfer, with the provinces insisting that only the cash portion should be counted while using the previous cost-shared transfers as their baseline for comparison [24].

To receive these federal funds, the provincial/territorial plans are required to adhere to the five national conditions specified in the 1984 Canada Health Act

(CHA) [25], which in turn are based on the earlier HIDS and Medical Care Act criteria. The national conditions require:

- *universal coverage*, with minor exceptions, all Canadian residents are classified as 'insured persons';
- *comprehensive coverage*, defined as covering all medically necessary services provided by hospitals or physicians ('insured services');
- *'reasonable access'* to insured services by insured persons, without user fees;
- provisions for *portability* across provinces for insured services; and
- ensuring that the publicly funded insurance plans, although not service delivery, be *publicly administered* to ensure accountability and lower administrative costs [26].

Within these general rules, provinces are free to make their own choices about how to organize and deliver care, and which services to cover.

Additional disputes have arisen from changes in how federal transfers would increase. The original formula called for EPF payments to be based on population (thus accepting that the federal per-capita payments would not reflect potential differences in health expenditures among various population subgroups), and to be indexed to inflation. To deal with the federal deficit, however, the federal government would unilaterally change the conditions, despite provincial objections. Thus, in 1986 indexing was reduced to GNP-2% (meaning that, as the tax take continued to grow with the economy, the cash portion would represent a disproportionate share of this reduction, and eventually might vanish altogether). In 1989, indexing was further reduced to GNP-3%, and in 1990 it was eliminated entirely. The resulting outcry eventually led to a new formula; in 1996, the federal government combined EPF with a remaining cost-shared program – the 1966 Canada Assistance Plan – which had covered some welfare programs (including some means-tested health care services which did not fall within the definition of 'insured services'). This allowed the federal government to cut the total transfer while still guaranteeing a 'cash floor'. The renamed 'Canada Health and Social Transfer' (CHST) retained the requirement that provinces comply with the national conditions for hospital and physician insurance as specified in the 1984 Canada Health Act, but did not impose any requirements for post-secondary education or welfare.

In effect, there has been no designated federal transfer for health since 1977, although all levels of government have often found it convenient to imply the contrary. For example, money flowing into provincial general revenues might be accompanied by an announcement that it was in support of medical equipment, but with no requirement that these resources not replace existing/announced funding for that purpose, provinces retained the ability to spend as they chose. Indeed, newspaper reporters observed that some provinces classified their purchases of lawn mowers or laser printers as coming under this medical equipment rubric [27]. In 2004, the CHST was again restructured; now, the separate pots were named the 'Canada Health Transfer' and the 'Canada Social Transfer'. This fiscal arrangement continues to allow considerable flexibility to the provinces, and to encourage constant dispute as to what should be counted (in both the numerator and the denominator). A recurring theme in Canadian political discourse is the vehement and unproductive 'shares debate', about what the federal contribution to health care actually is [28].

2.1.2
Delivering Health Care in Canada

The public administration criterion does not imply public delivery; rather, health care services are delivered, for the most part, by an array of private providers [29]. Hospitals are private, not-for-profit (albeit highly regulated) organizations, although individual hospital boards have been replaced in most provinces by quasipublic regional health authorities. Physicians remain private for-profit small businesses, although primary health care reform in most provinces has increasingly moved them towards various types of group practice. Private for-profit, investor-owned corporations predominate in such sectors as laboratory services and pharmaceuticals. This reliance on private, decentralized delivery gives the Canadian government fewer direct levers by which it can direct providers than would be the case in those countries with national health care systems. It also means that there is considerable variation both within and across jurisdictions. One size does not fit all, although there are key similarities. In turn, cost control efforts also varied.

2.1.3
Overall Expenditure Trends

Following the introduction of publicly funded universal coverage for both hospital and physician services, costs were relatively well controlled. In fact, the national data estimated that Canadian health expenditure per capita (in constant, 1997, dollars) increased by an average annual rate of 2.6% from 1975 to 1991 [9]. However, as noted above, a downturn in the economy and worry about deficits caused all levels of government to slam on the fiscal brakes [30–32]. Between 1991 and 1996, Canada was among the few OECD countries where health expenditure per capita actually declined, by an average rate of 0.3% per year. Subsequently, public dissatisfaction led to reinvestment [9]. As the OECD data note, between 2000 and 2005, health spending per capita increased in real terms by an average of 3.6%. This was lower than the OECD average (4.3% per year), but nonetheless represented a growing share of national income [2]. Indeed, total health expenditures in Canada were estimated at Can$ 131.4 billion in 2004 (Can$ 4109 per capita; $3161 in US$ PPP) and 9.8% of GDP [2], with forecasted increases of 6.4% (for 2006) and 6.6% (for 2007). After adjusting for inflation and population growth, this translated into real rates of increase of 3% and 3.2%, respectively [9]. Several additional indicators of trends over time, drawn from CIHI and OECD data, are listed in Table 2.1, to reflect the early stages of publicly funded care (1975), the period just before (1991) during (1997) and just after (1999) the major cost-containment efforts, and the most current data available from these sources.

Any glass can be seen as half-empty or half-full; Canadians do both with health care spending [7, 34]. As noted above, on the one hand, they stress that expenditures are high and growing and potentially unsustainable [3], although this conclusion is also challenged [13, 35, 36]. On the other hand, they argue that even more spending is needed [4, 37].

Table 2.1 Canada, trends over time, selected years 1975–2007.

	1975	1991	1999	Current
Population (millions)	23.0 (1976)	27.3	30.0 (2001)	33.0 (2007)
Life expectancy, total population at birth [33]	73.9 (1976)	77.8	79.0	80.2 (2004)
Hospitals: % of total health expenditures [9]	44.7%	38.8	29.8	28.4 (2007)
Hospitals: % of public sector health expenditures [9]	55.2	47.2	38.5	36.4 (2007)
Hospitals: current Can$ per capita [9]	235.70	917.41	882.38	1382.47 (2007)
Acute care beds in hospitals per 1000 population [33]	5.0 (1976)	4.0	3.3	2.9 (2004)
Practicing physicians per 1000 population [33]	1.7	2.1	2.1	2.2 (2005)
Practicing nurses per 1000 population [33]	9.1	11.2	10.1	10.0 (2005)
Pharmaceutical expenditure, % total [33]	8.9	11.8	15.5	17.8 (2006)

2.1.4
Cost Control Strategies Employed

As Ginsburg has noted for the US: "In reality, there are four basic options for slowing the trends in health care spending: one can increase the efficiency of health care delivery; increase the financial incentives for patients to limit their use of medical services; increase the administrative controls on the use of these services; or limit the resources available to the health care system. Health care systems throughout the world are pursuing variations on all four options, and the success of their efforts depends in part on how vigorously cost-containment tools are applied. But success does not come easily. For one thing, all health care spending represents someone else's income, and those who are facing a loss of income will work to block efforts to contain costs. In addition, each of these options, with the possible exception of the first, requires some people to get less medical care than they would like. For the most part, our leaders have been unwilling to acknowledge the inherent trade-offs between health care costs and people's access to care" [38]. The same observations apply to Canada.

2.2
Supply-Side Measures

Health costs are health incomes, and cost control accordingly has a major impact on providers [39]. An international review has confirmed that hard, enforceable budgets could indeed be successful in containing the expansion of health expenditures, although success was not automatic and might introduce other problems [40]. The Canadian example reinforces this conclusion. Government payers began with efforts to cap budgets for both hospital and physician services, but indeed found that these policies were difficult to sustain and led to an erosion of public satisfaction.

2.2.1
Capped Budgets

Because the vast majority of hospital and physician expenditures came from the single (provincial government) payers, it was relatively simple for provincial/territorial governments to constrain these budgets [26, 41]. Any method for paying providers has both strengths and weaknesses [42–44]. As Deber, Hollander and Jacobs [45] have noted, payment can be based on various combinations of: (i) the actual costs incurred; (ii) the time spent; (iii) the services provided; (iv) the population served and/or (v) outcomes achieved, either for individual patients and/or a population. Payment may flow to individual providers or to provider organizations. The research evidence strongly suggests that there is no single answer; rather, one must consider the incentives and disincentives inherent in alternative health care funding models [44, 45]. For example, the incentives of service-based funding approaches are to lower the cost per service (particularly if providers are allowed to retain the savings), to concentrate on the most lucrative services, and to increase volume (and often total costs) as long as the payment exceeds the cost. These incentives may be highly appropriate if policymakers wish to increase volumes (e.g. to improve access), but less desirable if they wish to encourage other modes of practice (e.g. fewer, longer visits). Similarly, global funding models may increase cost control but may lead to incentives to skimp on services (underservice) and/or select low-risk and low-need patients.

2.2.1.1 Hospitals

For budget capping, hospitals were the first, and easiest, target. Over 90% of hospital budgets came from publicly funded sources, and these were commonly paid through global budgets. From the perspective of easing implementation, capping budgets had the advantage of not directly attacking provider autonomy. Instead, local hospital boards were able to decide on their priorities. The budgetary squeeze resulted in a rapid reduction of the availability of hospital beds; provinces encouraged restructuring, which Iglehart [30] noted led to a drop from 1128 hospitals with 173 376 beds in 1991, to 877 hospitals with 122 006 beds in 1999, at the same time as the population increased. As shown in Table 2.1, in 1975 the hospitals had accounted for 55.2% of Canada's public sector health expenditures, and 44.7% of total expenditures. In 1991, it was still 47.2% of public sector health costs (38.8% of total), but by 1999, this had fallen to 38.5% of public expenditures (29.8% of total), and the decline continues (estimated at 36.4% of public, and 28.4% of total health expenditures for 2007) [9]. This also translated into an actual decrease in current dollars per capita for the 1994–1999 period, falling to Can$ 828.76 in 1997 before starting to increase again;

indeed, spending for hospitals did not catch up to the 1992 level until 2000. The OECD placed the number of acute-care hospital beds in Canada at 2.9 per 1000 population for 2004; although this exceeded the value in the US (2.7 in 2005), it was still lower than the OECD average of 3.9 [2].

These cost-control efforts were largely successful – as noted above, the per-capita, inflation-adjusted Canadian health expenditure actually fell during the mid 1990s. However, this had some (often unanticipated) consequences. The CHA required full coverage for medically necessary care if delivered in hospitals; however, this national requirement did not extend to services delivered by nonphysicians in the community. Although provinces could choose to go beyond these minimum requirements, there was considerable variability in what was covered. Capped budgets thus provided an opportunity – indeed, an incentive – for cost shifting, particularly for nursing care, rehabilitation and pharmaceuticals. It also left providers vulnerable, as hospitals often reacted to their budgetary constraints by laying off nurses and attempting to switch from full-time to part time/casual employment. One consequence was an exodus of nurses, which soon created a nursing shortage [46–52]. Constrained budgets and capacity reduction also meant that physicians had to battle for operating room time; those with less internal power (often the providers of elective surgery) often found themselves on the losing side. Ultimately, this would lead to waiting lists for those services, and strong pressure to add back resources, either inside or outside the publicly funded system, despite mixed international evidence that this approach was the most effective way of proceeding [53, 54].

2.2.1.2 Physicians

There was a similar effort to restrain physician costs, which was enormously complicated by the fact that most Canadian physicians were largely funded through fee-for-service [55, 56]. As leading health economists had noted, physician service expenditures could be viewed as the product of three sets of variables: the number of services; the price paid for each; and the mix of services. In turn, policy levers could be (and were) directed at the fees paid, at the volume of services, and/or at physician supply and physician incomes.

Provincial governments had relatively good control over the fees paid for particular services; about 99% of physician expenditures came from public sector sources, and in each province fee schedules were negotiated between the provincial government and the provincial medical association. However, this mechanism did not give the provinces control over the volume or mix of services provided. During the mid 1990s, the provinces all experimented with attempting to cap total physician payments. Such global expenditure caps give the payer predictable costs, but create a zero-sum game for providers. Individual providers can increase their income only at the expense of their colleagues. Eight of the 10 provinces attempted to incorporate 'hard' caps that called for recovering excess payments, usually by adjusting fee levels to ensure that the expenditure targets would be met [55]. Unsurprisingly, this evoked the usual dilemmas associated with common property resources ('tragedy of the commons'). One immediate response by physicians was to encourage controls over physician supply so that there would be fewer providers contending for shares of the now-limited pot [57] (this topic dealt with in Section 2.2.2.1). Another response was the intensification of arguments about the fees to be paid for particular procedures. The burden did not fall equally, however, and those practitioners less able to increase their volume of services were disproportionately affected. Physicians were also angered because data systems were insufficient to allow rapid response, which meant that the caps translated into retrospective clawbacks.

Shortages of particular services (especially elective care) soon materialized, the public objected, and the efforts at capping total physician payments were quickly abandoned [41, 58]. Instead, there has been a shift towards trying to encourage physicians to move from fee-for-service payment into various forms of team-based practice, with capitated funding taking on a greater role; this is progressing, albeit relatively slowly [59]. This approach in effect attempts to make physician services more closely resemble hospital services, giving payers more predictability of costs while retaining physician autonomy over the services they provide. In turn, the sense that capitated budgets gave incentives to underserve has led to more pressure for including pay for performance (P4P) elements in physician reimbursement [60].

2.2.1.3 Pharmaceuticals

Government efforts to restrain supply concentrated on the subsectors where they had monopsony power – particularly physicians and hospitals – and assumed that providers would in turn prioritize those services offering the greatest clinical benefit. Payers paid far less attention to controlling pharmaceutical costs, which arguably are currently the most out of control. Indeed, spending on pharmaceuticals has increased from 11.8% of total health spending in 1991, to 17.8% in 2006 [2, 3, 61]. However, much of this spending for pharmaceuticals came from private, often employment-based insurance, and significant gaps in coverage remain. Universal pharmaceutical coverage remains a missing piece of the coverage puzzle [62], while the fragmentation of payers has made cost control extremely difficult. Despite frequent calls for adding 'pharmacare' to coverage, governments have been reluctant to act, and there is widespread suspicion that costs are now too high to be able to do so, although some provinces have been expanding their coverage for certain populations [63–66]. Thus, although the CHA requires full coverage only for drugs delivered in hospitals, many provinces do provide outpatient pharmaceutical coverage for people with particular diseases (e.g. cancer, cystic fibrosis, HIV/AIDS), in selected age groups (e.g. those aged over 65 years), and/or in other vulnerable populations (e.g. those on social assistance, those receiving home care) (see http://www.hc-sc.gc.ca/hcs-sss/pharma/acces/ptprog_e.html for links to information about each province's plans). Here, too, there has been considerable flux. For example, British Columbia shifted from a policy providing coverage to seniors, to one employing means-testing [67]. In 1997, Quebec implemented a mixed system combining public and private payers. A recent analysis notes that the policy was successful at reducing access problems, but not at cost control; both, private and public expenditures for pharmaceuticals have soared (public expenditure by 286% over nine years, although the covered population grew by less than 3% during that time) [68]. Some provinces have experimented with cost control efforts – particularly British Columbia's efforts to employ reference-based pricing [69]. On occasion, provincial governments propose wider policy interventions, but quickly retreat once resistance from pharmaceutical companies, pharmacists and potentially affected patient groups appear [70, 71]. Like other OECD countries, Canada appears baffled by how best to control the costs of pharmaceuticals while still providing timely access to high-quality care, especially given the relative absence of levers in the current payment arrangements.

2.2.2
Health Human Resource Limits

Accompanying efforts to cap expenditures were those to control the supply of health human resources. It was recognized that health costs, to a large extent, were health incomes. Thus, to many policy makers, the providers were seen largely as cost drivers. Several approaches were employed to control health human resources.

2.2.2.1 Physicians
One set of policies protected those physicians already practicing by paying lower fees to new graduates. These policies were often linked to efforts to improve physician distribution (e.g. new physicians would be paid more if they practiced in areas deemed to be underserviced.) A related set of policies attempted to specify where physicians were allowed to practice through the requirement to have a billing number, until this was struck down by a court decision as infringing on the mobility rights of physicians [72, 73].

In order to practice, physicians must be licensed by their provincial regulatory colleges. Very soon after the enactment of universal coverage for physician services, the provincial colleges clamped down on foreign-trained physicians, making it more difficult for them to enter the system [74]. This policy has met with considerable resistance, particularly from those arguing that it is a waste of talent to have trained physicians driving taxi cabs (a widespread urban legend), and policies for dealing with International Medical Graduates (IMGs) are also in flux. Again, striking the appropriate balance between such policy goals as protecting the human rights of immigrants, ensuring quality, addressing geographical distribution of providers, and maintaining cost control has been neither simple nor uncontroversial [75, 76].

The provincial governments, in addition to having constitutional authority over education, also acted to reduce the number of trainees in medical schools. This policy resulted from a study on the need for physicians led by two leading health economists, Barer and Stoddart. These economists noted that enrollment had been increased in earlier years, in response to erroneous population projections, and concluded that there was now an oversupply [77]. Government responded by reducing enrollment in medical schools by 10% in 1992, in addition to the clamp down on foreign-trained physicians [30]. However, although Barer and Stoddart had assumed that much of the work being done by physicians could be done by other providers, the move towards health teams did not progress as quickly as anticipated; indeed, there was also a substantial drop in admissions to nursing schools [30]. Other assumptions they made also proved flawed, as governments implemented only some of their recommendations. Subsequent research clarified that the new cohort of physicians refused to work the extensive hours characteristic of earlier generations, which led to a change in the meaning and productivity – although not necessarily the reimbursement – of a Full Time Equivalent physician. New training regulations extended the time to produce family practitioners, and made it more difficult for physicians to retrain; spot shortages resulted. There was far less labor substitution than the report had recommended, and nurse practitioners came only slowly into the system [78]. Physicians worked differently, again confounding the projections [79]. In consequence, although the ratio of physicians to population in Canada compares favorably to that during the early 1980s, there is now widespread agreement among policy makers that there is a physician shortage which requires drastic – and expensive – remediation. Certainly, Canada's ratio of 2.2 practicing physicians to 1000 population in 2005 was considerably below the OECD average of 3.0 [2].

2.2.2.2 Other Health Care Professionals

Problems also arose among other health providers. To a large extent, this developed from the constraints on hospital global budgets, which translated into reductions in the number of nurses and other workers that they could employ. Although Canada still has more nurses (10.0 qualified nurses per 1000 population in 2005) than the OECD average of 8.6 [2], Canada reduced enrollment in nursing schools during the 1990s. This, combined with reductions in hospital employment, led to an actual decrease in the number of nurses per capita as nurses left the profession [52, 80]. Recognition that the profession was aging [81] again led to a policy reversal, and to increased attention to recruitment and retention issues. Again, this policy trend is not a new one – all OECD countries appear convinced that they have shortages of physicians and nurses, almost regardless of the ratios of providers to population [82].

A summary would thus conclude that these supply-side controls were very successful in constraining costs [83], but that they led to a backlash. This has been expressed in terms of declining public satisfaction, and heightened concern about access to care [4, 84]. In 1991, 61% of the Canadian public had rated their health care system as excellent or good. However, after several years of cost control this had dropped to 24%, although it has since rebounded very slowly. In consequence, the need to reinvest in health care began to dominate public discourse [30].

2.3
Organizational Reforms: Trade-Offs Between Competition and Cooperation

Another set of reforms concentrated on the need to change how care was delivered. Inherent contradictions existed between the reforms discussed here and in Section 2.4. One set of reforms spoke of the need to enhance cooperation among various components of health care [85, 86]. The usual dialogue spoke about the need to "... break down the silos" and create integrated models [87–89]. In Canada this took a variety of forms.

2.3.1
Cooperation: Regional Authorities

One set of reforms sought to replace individual providers with regional health authorities, to enable resources to be allocated most efficiently [90–94]. These authorities varied in their form, in the responsibilities given to them, and in their accomplishments. All provinces except Ontario implemented various forms of regionalization (although some have since reversed course), and Ontario is doing so at the time of writing, albeit with significant differences from other provinces. All included hospital services within the regional structures. However, whilst most provinces eliminated the formerly independent hospital boards, Ontario is retaining them and using the regional structures only to purchase services from provider organizations, in combination with detailed accountability agreements between the regions and the provinces, and the regions and the providers. It is noteworthy that none of these regional models has yet included physicians, thereby accentuating difficulties in integrating medical services with other subsectors.

2.3.2
Cooperation: Primary Care Reform

Another set of reforms sought to reorganize the delivery of primary health care, again to encourage cooperation and service integration, and sparked in part by difficulties in finding primary care physicians in some parts of the country. This usually has taken the form of encouraging physicians to move from solo-practice (usually funded fee-for-service) to a variety of group-practice arrangements, often including other health care providers, and funded at least in part through capitation. Funders also encouraged a move towards defining which patients would be served in which practices, often through rostering [88, 95]. These reforms are progressing slowly, but steadily, with an increasing proportion of physicians working in alternative practice arrangements.

2.3.3
Waiting Time Strategies

Another set of reforms has attempted to deal with a major negative side effect of capacity controls, namely increased waiting times [53, 54, 96–100]. Capped global budgets for hospitals encouraged them to prioritize their services but, by definition, such processes created winners and losers. Those providing elective surgery proved particularly susceptible to constraints, and waiting lists resulted, particularly for such elective services as hip/knee replacements, cataract surgery and nonurgent diagnostic imaging. One set of responses has sought more resources, either within hospitals, or within private clinics [29]. Another, however, recognized that better system management (particularly improving logistics) gave the potential for 'win-win' situations which could improve access, improve patient satisfaction and affect waiting times [101, 102]. In consequence, there have been efforts to implement waiting list management strategies; these contain elements of cooperation, competition and appropriateness.

Among the most contentious have been the efforts to coordinate formerly autonomous providers and create actual systems of care. This involves encouraging a single point of entry, prioritizing cases, setting targets and indicators, and ensuring that these targets could be met. The federal/provincial governments agreed in 2004 on a Ten-Year Plan which agreed to invest Can$ 4.5 billion over six years to reduce waiting times for five priority areas: cancer, heart disease, diagnostic imaging, joint replacement, and sight restoration [103]. The success of these efforts has been variable. Because the first step was attempting to define targets and indicators, they are also linked to the 'appropriateness' strategies discussed below. Unsurprisingly, these approaches have been most successful for services where appropriateness can be defined by clinical experts (e.g. cardiac surgery, cancer care), and relatively less successful in dealing with services where 'need' has proven to be more ambiguous. Indeed, some of these services resemble a treadmill; the provision of more resources, rather than eliminating waiting times, has instead encouraged more cases to be recommended for services, leaving the waiting lists relatively unaffected. Nonetheless, in the view of most observers these ongoing activities have enormous potential in encouraging better care [96, 104].

2.3.4
Competition

In contrast, another set of reforms sought to encourage competition rather than cooperation, in the belief that this might harness market forces and thus bring about cost control. The primary use has been in areas that do not fall under the CHA requirements, and hence where mixed financing exists. The province of Ontario, for example, has used managed competition for home care, in which a series of arm's-length regional organizations (named Community Care Access Centres) were given the responsibility of purchasing home care services at 'best quality, best price'. The results have been mixed; costs for services sufficiently specialized as to discourage competition have, if anything, increased [105–108]. Few other provinces have followed suit [109], although there has been increasing pressure to shift to service-based funding and allow public and private providers to compete for publicly paid services.

A related set of efforts is examining whether there might be a role for increased use of service-based funding, which in turn might allow competition among providers [110]. Thus, there is considerable pressure to move hospitals from the current global budgets to service-based funding. Similarly, there are some tentative moves towards P4P for physician services, in which payers use financial levers to encourage health care providers to achieve measured standards of patient care [111, 112]. P4P exists within the context of how providers are paid, and the incentives associated with this policy instrument.

Again, these are in the early stages, particularly because the implications of moving towards service-based funding for cost control are by no means clear; in fact, they may erode the relative success in cost control that had been achieved in the physician and hospital sectors. However, it is noteworthy that waiting lists have been managed both through improved cooperation/management, and through providing additional service-based funding for specified procedures [96].

2.4
Demand-Side Measures

2.4.1
Technology Assessment/Appropriateness

To date, the major efforts on the demand side have been efforts to increase the appropriateness of the care provided. Recognizing that unneeded care may represent cost without benefit, or even cost to cause harm [113, 114], providers and payers have been moving towards an increased use of such approaches as evidence-based medicine, technology assessment and various forms of clinical guidelines [115]. One example – trying to develop appropriateness criteria to guide waiting lists – has been mentioned above. The provinces have moved, albeit very slowly, towards the use of a common drug review process to give advice about which pharmaceuticals should be added to hospital formularies, provincial drug plans and/or private drug plans [116–121]. This effort has to contend with the structural barriers to any national approach, and hence rely upon voluntary cooperation among the various players in working together, and voluntary agreement to abide by the results of the processes. Neither has been simple to achieve, but efforts continue.

At the national level, the Canadian Agency for Drugs and Technologies in Health (CADTH) is a national not-for-profit organization, originally named the Canadian Coordinating Office for Health Technology Assessment (CCOHTA). It was established in 1990, and renamed in 2006, and given a mandate to provide decision makers at the federal, provincial and territorial levels "…with credible, impartial advice and evidence-based information about the effectiveness and efficiency of drugs and other health technologies". (The activities of the CADTH are described on their web site: http://www.cadth.ca/.) The CADTH is responsible for managing the above-mentioned Common Drug Review, as well as for programs for health technology assessment and identifying and promoting optimal drug therapy. Although the Agency has produced a number of reports, it has a purely advisory role. Even here, federal–provincial issues intrude; for example, Quebec is not a member but has its own body, the AETMIS (the Quebec Agency for Health Services and Technology Assessment). Formal technology assessment activities also occur in other provinces, often in university or other 'think-tanks', and also within provincial governments [122]. For example, Alberta has worked with local universities and with the Institute of Health Economics (IHE) to conduct and implement a number of such assessments [123]. These bodies are confronted with the ongoing challenges faced by any technology assessment – including how to value costs and benefits, whose costs and consequences matter [124] – coupled with very few levers to translate their recommendations into policy. Nonetheless, they play a potentially important role, particularly in the long term.

2.4.2
Population Health

Finally, there has been recognition that the health of a population is the function of many things other than health care. The Lalonde Report [125] was a pioneer in recognizing the importance of various determinants of health; it attributed health status to four factors: medical services, genetic, lifestyle, and environmental issues. This recognition of the importance of health promotion and disease prevention evoked considerable attention, initially concentrating upon changing personal lifestyle habits to decrease risk factors such as smoking, obesity and physical inactivity, but over time also stressing broader social factors, including poverty and inequity [126]. Not surprisingly, this emphasis on the health of populations has proved difficult to fully implement (although smoking rates have indeed decreased considerably) [127, 128]. There is considerable regional variation in these lifestyle and environmental factors, leading to variations in health outcomes across the country [75, 129]. Canada has encouraged various health-promotion/disease-prevention activities, but again this is complicated by jurisdictional issues and by the fact that many key determinants of health are not under the control of ministries of health. Air quality, for example, is heavily dependent upon industrial and transportation policy. Nonetheless, Canada has seen strong programs in such areas as immunization (with the federal government acting to pay for certain vaccines); efforts to discourage obesity [130]; improved food labeling (*trans*-fats, sodium, etc.) to help people voluntarily improve their diets, and so on. Among the greater success stories is smoking cessation, with Canada's smoking rates having decreased considerably [2].

2.5
The Current Situation

An assessment of polling data concludes: "Herein lies one of the puzzles of Canadian health care: Canadians increasingly view the health care system as unsustainable and under threat, even as their own experiences with the system are mostly positive" [4]. The current picture has not changed for decades – Canadian health care is popular, successful, and at major risk. Canadian efforts to achieve high-quality, timely care at an affordable, sustainable cost have evoked a series of commissions and reports [88, 89, 131, 132]. Their recommendations have been largely ignored, and the debate continues to return to questions which academics had hoped, vainly, had been settled.

At present, cost control is far lower on the agenda than are the access issues. The major approach to improving access has been the call for adding resources. In the longer term, this is likely to lead to another effort to curb expenditures, particularly whenever the inevitable economic slowdown materializes. One pervasive policy difficulty is that governments have ongoing conflict of interests. As stewards of the well-being of the country, it is clearly advantageous to ensure that policy minimizes the burden on the economy while maintaining a high-quality health care system that provides timely access to needed care. From the viewpoint of governments as single payers, however, it is equally clearly advantageous to shift those costs off government budgets, leaving room for other spending and/or tax cuts. The main restraint against that second tendency has been the national conditions inherent in the Canada Health Act.

Economists will note that the strategies discussed have not emphasized approaches that attempt to influence behavior of patients via copayments/user fees, although there are continual efforts by some policy advocates to bring this set of policy levers into play. Any such suggestions incur immediate opposition and cries of 'two-tier medicine'. The CHA was designed to act as a barrier to the use of these levers for insured services (doctors, hospitals), and has successfully done so. However, considerable use is already made of such copayments and coverage limitations for services outside Medicare, including outpatient pharmaceuticals, long-term care, home care, dentistry and rehabilitation services, and Canada has among the highest proportion of out-of-pocket costs among the OECD countries [133]. Cost escalation has been highest in these mixed-funding areas, particularly pharmaceuticals. Any weakening of the CHA, or the ability to enforce it, appears likely also to lead to a greater reliance on cost-shifting for physician and hospital services.

Here lies one key threat to the current system. To the extent that demands to disentangle the federal government from areas under provincial jurisdiction trump the desire for national levels of services, the CHA restraints may erode. At present, the Conservative government of Stephen Harper is on record as advocating such disentanglement, but as a minority government has not yet acted on this policy. Instituting regulations to ensure that the federal government cannot intervene in such areas is also a policy plank of the Quebec-based separatist party, the Bloc Quebecois. There are few strong voices speaking for a federal role, and the public has not connected these seemingly arcane questions about federal and provincial roles with the fate of health care. Should a subsequent government choose to act, it has the ability to use some potential levers which cannot be undone, particularly replacing the remaining cash transfers by tax points. Should this happen, the CHA will become unenforceable, and each province will be free to go its own way, including privatizing the system.

Another potential lever is the move towards service-based funding, which would simplify a move towards allowing competition for delivery. To the extent that risk adjustment is complex, this may also open the way towards privatization and the erosion of less-profitable (but still important) services, as well as potential access difficulties for less-profitable clients.

Nonetheless, Canada is famous for muddling through. This chapter concludes with the hope that it continue to do so, and that the next swing of the pendulum will emphasize quality and appropriateness as its mechanism for ensuring cost containment without sacrificing efficiency, or universal coverage.

References

1 Docteur, E. and Oxley, H. (2003) Health-Care Systems: Lessons from the Reform Experience, OECD Health Working Papers, No. 9, OECD Publishing. Downloaded from http://www.oecd.org/dataoecd/5/53/22364122.pdf

2 Organisation for Economic Co-operation and Development (OECD) (2007) OECD Health Data 2008: how does Canada compare, Organisation for Economic Co-operation and Development, Paris, France. Downloaded from http://www.oecd.org/dataoecd/46/33/38979719.pdf (accessed 16 May 2008).

3 Canadian Institute for Health Information (2007) Health care in Canada 2007, Canadian Institute for Health Information, Ottawa, Canada. Downloaded from http://secure.cihi.ca/cihiweb/products/hcic2007_e.pdf (accessed 16 May 2008).

4 Soroka, S.N. (2007) Canadian perceptions of the health care system. A report to the Health Council of Canada, Toronto, Canada: Health Council of Canada, February. Downloaded from http://www.healthcouncilcanada.ca/docs/rpts/2007/Public%20Perceptions%20-%20English%20Final_Feb-07.pdf (accessed 16 May 2008).

5 Schoen, C., Osborn, R., Doty, M.M., Bishop, M., Peugh, J. and Murukutla, N. (2007) Toward higher-performance health systems: adults' health care experiences in seven countries. *Health Affairs*, **26** (6), w717–w724.

6 Deber, R. and Swan, B. (1999) Canadian health expenditures: where do we really stand internationally? *Canadian Medical Association Journal*, **160** (12), 1730–34.

7 Marchildon, G.P., McIntosh, T. and Forest, P.-G. (2004) *The Fiscal Sustainability of Health Care in Canada: Romanow Papers*, Vol. 1, University of Toronto Press, Toronto, Canada.

8 Harris, M. and Manning, P. (2006) Rebalanced and revitalized: a Canada strong and free, The Fraser Institute, Vancouver, B.C., June. Downloaded from http://www.fraserinstitute.org/commerce.web/product_files/RebalancedRevitalized.pdf (accessed 16 May 2008).

9 Canadian Institute for Health Information (2007) National Health Expenditure Trends, 1975–2007, Canadian Institute for Health Information, Ottawa, Canada. Downloaded from http://secure.cihi.ca/cihiweb/products/NHET_1975_2007_e.pdf

10 Marchildon, G.P. (2006) *Health Care Systems in Transition*, 1st edn, University of Toronto Press, Toronto, Canada. Downloaded from http://www.euro.who.int/document/e87954.pdf

11 Maioni, A. (2004) Roles and responsibilities in health care policy, in *The Governance of Health Care in Canada: Romanow Papers*, Vol. 3 (eds T. McIntosh, P.-G. Forest and G.P. Marchildon), University of Toronto Press, Toronto, Canada, pp. 169–198.

12 Rode, M. and Rushton, M. (2004) Increasing provincial revenues for health care, in *The Fiscal Sustainability of Health Care in Canada: Romanow Papers*, Vol. 1 (eds G.P. Marchildon, T. McIntosh and P.-G. Forest), University of Toronto Press, Toronto, Canada, pp. 299–319.

13 Boychuk, G.W. (2004) The changing political and economic environment of health care, in *The Fiscal Sustainability of Health Care in Canada: Romanow Papers*, Vol. 1 (eds G.P. Marchildon, T. McIntosh and P.-G. Forest), University of Toronto Press, Toronto, Canada, pp. 320–339.

14 Fierlbeck, K. (2004) Paying to play? Government financing and agenda setting for health care, in *The Fiscal Sustainability of Health Care in Canada: Romanow Papers*, Vol. 1 (eds G.P.

Marchildon, T. McIntosh and P.-G.
Forest), University of Toronto Press,
Toronto, Canada, pp. 340–365.

15 Deber, R. (2000) Getting what we pay for
myths and realities about financing
Canada's health care system. *Health Law
in Canada*, **21** (2), 9–56.

16 Deber, R. (2003) Health care reform:
lessons from Canada. *American Journal of
Public Health*, **93** (1), 20–24.

17 Deber, R. (2003) Why did the World Health
Organization rate Canada's health system
as 30th? *Longwoods Review*, **2** (1), 2–7.

18 Deber, R., Millan, K., Shapiro, H. and
McDougall, C.W. (2006) A cautionary tale
of downloading public health in Ontario:
what does it say about the need for national
standards for more than doctors and
hospitals? *Healthcare Policy*, **2** (2), 56–71.

19 McIntosh, T. (2004) Intergovernmental
relations, social policy and federal
transfers after Romanow. *Canadian Public
Administration*, **47** (1), 27–51.

20 Banting, K.G., Brown, D.M. and
Courchene, T.J. (1994) The Future of
Fiscal Federalism, Queens University
School of Policy Studies, Kingston.

21 Taylor, M.G. (1987) *Health Insurance and
Canadian Public Policy, The Seven Decisions
that Created the Canadian Health
Insurance System and Their Outcomes*, 2nd
edn, McGill-Queen's University Press,
Kingston.

22 Department of Finance Canada (2007)
A brief history of the health and social
transfers. Department of Finance Canada,
March. Downloaded from http://www.
fin.gc.ca/FEDPROV/hise.html (accessed
16 May 2008).

23 Deber, R. (1996) National standards
in health care. *Policy Options*, **17** (5),
43–45.

24 Van Loon, R.J. (1978) From shared cost to
block funding and beyond: the politics of
health insurance in Canada. *Journal of
Health Politics, Policy and Law*, **2** (4),
454–478.

25 Government of Canada. *Canada Health
Act Bill C-3*.Statutes of Canada. 1984,

32–3. Elizabeth II (R.S.C. 1985, c. 6; R.S.
C. 1989, c. C-6) (1984).

26 Woolhandler, S., Campbell, T. and
Himmelstein, D.U. (2003) Costs of health
care administration in the United States
and Canada. *New England Journal of
Medicine*, **349** (8), 768–775.

27 Laurent, S. and Vaillancourt, F. (2004)
Federal-Provincial Transfers for Social
Programs in Canada: their Status in
May 2004, IRPP Working Paper Series,
no. 2004–07, Institute for Research
on Public Policy, Montréal, Canada.
Downloaded from http://www.irpp.org/
wp/archive/wp2004-07.pdf (accessed
16 May 2008).

28 Deber, R. (2000) Who wants to pay for
health care? *Canadian Medical Association
Journal*, **163** (1), 43–44.

29 Deber, R. (2004) Delivering health
care services: public, not-for-profit, or
private? in *The Fiscal Sustainability of
Health Care in Canada: Romanow Papers*,
Vol. 1 (eds G.P. Marchildon, T. McIntosh
and P.-G. Forest), University of
Toronto Press, Toronto, Canada,
pp. 233–296.

30 Iglehart, J.K. (2000) Revisiting the
Canadian health care system. *New
England Journal of Medicine*, **342** (26),
2007–2012.

31 Detsky, A.S. and Naylor, D. (2003)
Canada's health care system – reform
delayed. *New England Journal of Medicine*,
349 (8), 804–810.

32 Brimacombe, G.G. (2002) Every number
tells a story: a review of public and private
health expenditures and revenues in
Canada, 1980–2000, The Conference
Board of Canada, Ottawa, Canada.

33 Organisation for Economic Co-operation
and Development (2007) Health at a
glance 2007: OECD Indicators,
Organisation for Economic Co-operation
and Development, Paris, France.
Downloaded from http://lysander.
sourceoecd.org/vl=9846750/cl=11/
nw=1/rpsv/ij/oecdthemes/99980142/
v2007n23/s1/pl.

34 Stuart, N. and Adams, J. (2007) The sustainability of Canada's healthcare system: a framework for advancing the debate. *Longwoods Review,* **4** (4), 96–102.

35 Evans, R.G. (2004) Financing health care: options, consequences, and objectives, in *The Fiscal Sustainability of Health Care in Canada: Romanow Papers,* Vol. 1 (eds G.P. Marchildon, T. McIntosh and P.-G. Forest), University of Toronto Press, Toronto, Canada. pp. 139–196.

36 Evans, R.G. (2007) Economic myths and political realities: the inequality agenda and the sustainability of Medicare. Working paper, UBC Centre for Health Services and Policy Research, University of British Columbia, Vancouver, Canada, July. Downloaded from http://www.chspr.ubc.ca/files/publications/2007/chspr07-13W.pdf (accessed 16 May 2008).

37 Mendelsohn, M.(2002) Canadians' thoughts on their health care system: preserving the Canadian model through innovation. Paper prepared for the Commission on the Future of Health Care in Canada, June.

38 Ginsburg, P.B. (2004) Controlling health care costs. *New England Journal of Medicine,* **351** (16), 1591–1593.

39 Bodenheimer, T. (2005) High and rising health care costs: Part 3: the role of health care providers. *Annals of Internal Medicine,* **142**, 996–1002.

40 Carrin, G. and Hanvoravongchai, P. (2003) Provider payments and patient charges as policy tools for cost-containment: how successful are they in high-income countries? *Human Resources for Health,* **1** (6), 1–10.

41 Evans, R.G., Lomas, J., Barer, M.L., Labelle, R.J., Fooks, C., Stoddart, G.L. *et al.* (1989) Controlling health expenditures – the Canadian reality. *New England Journal of Medicine,* **320** (9), 571–577.

42 Robinson, J.C. (2001) Theory and practice in the design of physician payment incentives. *The Milbank Quarterly,* **79** (2), 149–177.

43 Grignon, M., Paris, V. and Polton, D. (2004) The influence of physician payment methods on the efficiency of the health care system, in *Changing Health Care in Canada: Romanow Papers,* Vol. 2 (eds P.-G. Forest, G.P. Marchildon and T. McIntosh), University of Toronto Press, Toronto, Canada, pp. 205–239.

44 Devlin, R.A., Sarma, S. and Hogg, W. (2006) Remunerating primary care physicians: Emerging directions and policy options for Canada. *Healthcare Quarterly,* **9** (3), 34–42.

45 Deber, R., Hollander, M.J. and Jacobs, P. (2008) Models of funding and reimbursement in health care: a conceptual framework. *Canadian Public Administration,* **51** (3), 381–405.

46 Canadian Institute for Health Information (2007) Workforce trends of registered nurses in Canada, 2006. Registered Nurses Database: Ottawa, ON: CIHI. Downloaded from http://secure.cihi.ca/cihiweb/products/workforce_trends_of_rns_2006_e.pdf (accessed 16 May 2008).

47 O'Brien-Pallas, L., Tomblin Murphy, G., White, S., Hayes, L., Baumann, A., Higgin, A. *et al.* (2005) Building the future: an integrated strategy for nursing human resources in Canada: research synthesis report, The Nursing Sector Study Corporation, Ottawa, Canada, May. Downloaded from http://www.hhrchair.ca/images/CMSImages/Building%20the%20future%20Research-Synthesis-Report.pdf (accessed 16 May 2008).

48 Med-Emerg Inc. (2006) Building the future: an integrated strategy for nursing human resources in Canada: phase II final report. The Nursing Sector Study Corporation, Ottawa, ON, May. Downloaded from http://www.cna-nurses.ca/CNA/documents/pdf/publications/Phase_II_Final_Report_e.pdf (accessed 16 May 2008).

49 Vujicic, M. and Evans, R.G. (2005) The impact of deficit reduction on the nursing labour market in Canada. *Applied Health Economics and Health Policy*, **4** (2), 99–110.

50 Baumann, A., Giovannetti, P., O'Brien-Pallas, L., Mallette, C., Deber, R., Blythe, J. *et al.* (2001) Healthcare restructuring: the impact of job change. *Canadian Journal of Nursing Leadership*, **14** (1), 14–20.

51 Kazanjian, A. (2000) Changes in the nursing workforce and policy implications, Nursing Workforce Study, Vol. V, Health Human Resources Unit, Centre for Health Services and Policy Research, University of British Columbia, HHRU.00: 7. Downloaded from http://www.chspr.ubc.ca/files/publications/2000/hhru00-07_NWPv5.pdf

52 Alameddine, M., Laporte, A., Baumann, A., O'Brien-Pallas, L., Croxford, R., Mildon, B. *et al.* (2006) Where are nurses working? Employment patterns by sub-sector in Ontario, Canada. *Healthcare Policy*, **1** (3), 65–86.

53 Hurst, J. and Siciliani, L. (2005) Tackling excessive waiting times for elective surgery: a comparison of policies in twelve OECD countries. *Health Policy*, **72** (2), 201–215.

54 Sanmartin, C., Shortt, S.E.D., Barer, M.L., Sheps, S., Lewis, S. and McDonald, P.W. (2000) Waiting for medical services in Canada: lots of heat, but little light. *Canadian Medical Association Journal*, **162** (9), 1305–1310.

55 Barer, M.L., Lomas, J. and Sanmartin, C. (1996) Re-minding our Ps and Qs: medical cost controls in Canada. *Health Affairs*, **15** (2), 216–234.

56 Barer, M.L., Evans, R.G. and Labelle, R.J. (1988) Fee controls as cost control: tales from the frozen north. *The Milbank Quarterly*, **66** (1), 1–64.

57 Hurley, J., Lomas, J. and Goldsmith, L.J. (1997) Physician responses to global physician expenditure budgets in Canada: a common property perspective. *The Milbank Quarterly*, **75** (3), 343–364.

58 Barer, M.L. and Evans, R.G. (1986) Riding north on a south-bound horse?

Expenditures, prices, utilization and incomes in the Canadian health care system, in *Medicare at Maturity: Achievements Lessons and Challenges* (eds R.G. Evans and G.L. Stoddart), University of Calgary Press, Calgary, pp. 53–163.

59 Hutchison, B., Abelson, J. and Lavis, J. (2001) Primary care in Canada: so much innovation, so little change. *Health Affairs*, **20** (3), 116–131.

60 Bell, C.M. and Levinson, W. (2007) Pay for performance: learning about quality. *Canadian Medical Association Journal*, **176** (12), 1717–1719.

61 Paris, V. and Docteur, E. (2006) Pharmaceutical pricing and reimbursement policies in Canada. OECD Health Working Papers, 24, Organisation for Economic Co-operation and Development, Paris, France, 22 December. Downloaded from http://www.oecd.org/dataoecd/21/40/37868186.pdf (accessed 16 May 2008).

62 Health Council of Canada (2005) Health care renewal in Canada: accelerating change. Toronto, ON: Health Council of Canada, January. Downloaded from http://www.healthcouncilcanada.ca/docs/rpts/2005/Accelerating_Change_HCC_2005.pdf (accessed 16 May 2008).

63 Evans, R.G. (2005) Baneful legacy: Medicare and Mr Trudeau. *Healthcare Policy*, **1** (1), 20–25.

64 Willison, D., Grootendorst, P.V. and Hurley, J. (1998) Variation in Pharmacare Coverage across Canada, Centre for Health Economics and Policy Analysis Research Working Paper 98-08, McMaster University, Hamilton. Downloaded from http://www.chepa.org/portals/0/pdf/98-08.pdf

65 Canadian Health Services Research Foundation (2002) Pharmacare in Canada. Ottawa: Commission on the Future of Health Care in Canada, Discussion Paper prepared for the Commission on the Future of Health Care in Canada, May. Downloaded from

http://www.chsrf.ca/other_documents/
romanow/pdf/pharmacare_e.pdf

66 Sketris, I.S., Brown, M.G. and Murphy,
A.L. (2004) Policy choices for pharmacare:
the need to examine benefit design,
medication management strategies and
evaluation. *Healthcare Papers*, 4 (3), 36–45.

67 Morgan, S., Evans, R.G., Hanley, G.E.,
Caetano, P.A. and Black, C. (2006)
Income-based drug coverage in British
Columbia: lessons for BC and the rest
of Canada. *Healthcare Policy*, 2 (2),
115–127.

68 Pomey, M-P., Forest, P.-G., Palley, H.A.
and Martin, E. (2007) Public/private
partnerships for prescription drug
coverage: policy formulation and
outcomes in Quebec's universal drug
insurance program, with comparisons to
the Medicare prescription drug program
in the United States. *The Milbank
Quarterly*, 85 (3), 469–498.

69 Morgan, S.G., Barer, M.L. and Agnew,
J.D. (2003) Whither seniors' pharmacare:
Lessons from (and for) Canada. *Health
Affairs*, 22 (3), 49–59.

70 Deber, R. (2004) Taking our medicine:
who should pay for what? *Healthcare
Papers*, 4 (3), 27–28.

71 Willison, D., Wiktorowicz, M.,
Grootendorst, P., O'Brien, B., Levine, M.,
Deber, R. *et al.* (2001) International
experience with pharmaceutical policy:
common challenges and lessons for
Canada. Paper 01-8, CHEPA Working
Paper Series. Centre for Health
Economics and Policy Analysis (CHEPA),
McMaster University Hamilton, Canada.
Downloaded from http://www.chepa.org/
portals/0/pdf/01-08.pdf (accessed 16 May
2008).

72 Deber, R. and Heiber, S. (1988) Freedom,
equality and the Charter of Rights:
regulating physician reimbursement.
Canadian Public Administration, 31 (4),
566–589.

73 Manfredi, C.P. and Maioni, A. (2002)
Courts and health policy: judicial policy
making and publicly funded health care in

Canada. *Journal of Health Politics, Policy
and Law*, 27 (2), 213–240.

74 Evans, R.G. (1976) Does Canada have too
many doctors? Why nobody loves an
immigrant physician. *Canadian Public
Policy*, 2 (2), 147–160.

75 Health Council of Canada (2007) Health
care renewal in Canada: measuring up?
Annual report to Canadians 2006. Health
Council of Canada, Toronto, Canada.
February. Downloaded from http://www.
healthcouncilcanada.ca/docs/rpts/2007/
HCC_MeasuringUp_2007ENG.pdf
(accessed 16 May 2008).

76 (a) Canadian Institute for Health
Information (2006) Health personnel
trends in Canada, 1995 to 2004 (Revised
July 2006). Downloaded from http://
secure.cihi.ca/cihiweb/products/
Health_Personnel_Trend_1995-2004_e.
pdf (accessed 16 May 2008).

77 Barer, M.L. and Stoddart, G.L. (1991)
Toward integrated medical resource
policies for Canada: background
document. Report prepared for the
Federal/Provincial/Territorial Conference
of Deputy Ministers of Health, June.
Downloaded from http://
www.chspr.ubc.ca/files/publications/
1991/hpru91-06D.pdf

78 Chan, B.T.B. (2002) From perceived
surplus to perceived shortage: what
happened to Canada's physician
workforce in the 1990s? Canadian
Institute for Health Information, Toronto,
Canada, June. Downloaded from http://
secure.cihi.ca/cihiweb/products/chan-
jun02.pdf (accessed 16 May 2008).

79 Tepper, J. (2004) The Evolving Role of
Canada's Family Physicians: 1992–2001,
Canadian Institute for Health Information, Ottawa, Canada. Downloaded from
http://secure.cihi.ca/cihiweb/products/
physiciansREPORT_erg.pdf

80 Baumann, A.O., O'Brien-Pallas, L.L.,
Deber, R., Donner, G., Semogas, D. and
Silverman, B. (1996) Downsizing in the
hospital system: a restructuring process.
Healthcare Management Forum, 9 (4), 5–23.

81 O'Brien-Pallas, L., Alksnis, C. and Wang, S. (2003) Bringing the Future into Focus: Projecting RN Retirement in Canada, Canadian Institute for Health Information, Ottawa, Canada. Downloaded from http://secure.cihi.ca/cihiweb/products/RNRetirement2003_e.pdf

82 Simoens, S., Villeneuve, M. and Hurst, J. (2005) Tackling nurse shortages in OECD countries. Paris, France: OECD Health Working Papers no. 19. Downloaded from http://www.oecd.org/dataoecd/11/10/34571365.pdf

83 Evans, R.G., Barer, M.L. and Hertzman, C. (1991) The 20-year experiment: accounting for, explaining, and evaluating health care cost containment in Canada and the United States. *Annual Review of Public Health*, **12**, 481–518.

84 Blendon, R.J., Schoen, C., DesRoches, C., Osborn, R. and Zapert, K. (2003) Common concerns amid diverse systems: Health care experiences in five countries. *Health Affairs*, **22** (3), 106–121.

85 Shortell, S.M., Kellogg, J.L., Anderson, D.A., Mitchell, J.B. and Morgan, K.L. (1993) Creating organized delivery systems: the barriers and facilitators. *Hospital & Health Services Administration*, **38** (4), 447–466.

86 Boerma, W.G.W. (2006) Coordination and integration in European primary care, in *Primary Care in the Driver's Seat?* (eds R.B. Saltman, A. Rico and W. Boerma), Open University Press, Berkshire, England, pp. 3–21.

87 Denis, J.-L. (2004) Governance and management of change in Canada's health system, in *Changing Health Care in Canada: Romanow Papers*, Vol. 2 (eds P.-G. Forest, G.P. Marchildon and T. McIntosh), University of Toronto Press, Toronto, Canada, pp. 82–114.

88 Canada. Parliament. Standing senate committee on social Affairs, Science and Technology (2002) The health of Canadians: the federal role. Final report volume six: recommendations for reform. Ottawa, ON: Interim report on the state of the health care system in Canada. Downloaded from http://www.parl.gc.ca/37/2/parlbus/commbus/senate/com-e/SOCI-E/rep-e/repoct02vol6-e.htm (accessed 16 May 2008).

89 Commission on the future of health care in Canada (2002) Building on values: the future of health care in Canada: final report. Ottawa, Canada: Queen's Printer, 28 November. Downloaded from http://www.hc-sc.gc.ca/hcs-sss/alt_formats/hpb-dgps/pdf/hhr/romanow-erg.pdf (accessed 16 May 2008) (see also http://www.hc-sc.gc.ca/english/pdf/romanow/pdfs/HCC_Final_Report.pdf).

90 Marchildon, G.P. (2005) Health systems in transition: Canada. WHO Regional Office for Europe on behalf of the European Observatory on Health Systems and Policies. Copenhagen, Denmark. Downloaded from http://www.euro.who.int/Document/E87954.pdf

91 Lewis, S. and Kouri, D. (2004) Regionalization: making sense of the Canadian experience. *Healthcare Papers*, **5** (1), 12–31.

92 Church, J. and Barker, P. (1998) Regionalization of health services in Canada: a critical perspective. *International Journal of Health Services*, **28** (3), 467–486.

93 Dorland, J.L. and Davis, S.M. (1996) How Many Roads . . .? Proceedings of The Queen's-CMA Conference on Regionalization and Decentralization in Health Care, Queen's University School of Policy Studies, Kingston.

94 Conrad, P.A. (2007) Do regional models matter? Resource allocation to home care in the maritime provinces (Dissertation). University of Toronto, Toronto, Canada.

95 Health Council of Canada (2005) Primary health care. A background paper to accompany Health Care Renewal in Canada: Accelerating Change, Toronto, Canada. Downloaded from http://www.healthcouncilcanada.ca/docs/papers/2005/BkgrdPrimaryCareENG.pdf

96 Health Council of Canada (2007) Wading through wait times: what do meaningful reductions and guarantees mean? An update on wait times for health care. Health Council of Canada, Toronto, Canada, June. Downloaded from http://healthcouncilcanada.ca/ docs/rpts/2007/wait_times/hcc_wait-times-update_200706_FINAL% 20ENGLISH.pdf (accessed 16 May 2008).

97 Lewis, S., Barer, M.L., Sanmartin, C., Sheps, S., Shortt, S.E.D. and McDonald, P.W. (2000) Ending waiting-list mismanagement: principles and practice. *Canadian Medical Association Journal*, **162** (9), 1297–1300.

98 Appleby, J., Boyle, S., Devlin, N., Harley, M., Harrison, A. and Locock, L. (2003) Sustaining reductions in waiting times: identifying successful strategies. London, England: King's Fund. Downloaded from http://www.kingsfund.org.uk/ publications/kings_fund_publications/ sustaining.html (last accessed 5 January 2007).

99 Canadian Institute for Health Information, Statistics, Canada (2003) Health care in Canada, 2003. CIHI, Statistics Canada. Ottawa, Canada. Downloaded from http://secure.cihi.ca/ cihiweb/products/hcic2003_e.pdf (accessed 16 May 2008).

100 Sanmartin, C., Gendron, F., Berthelot, J.-M. and Murphy, K. (2004) Access to health care services in Canada, 2003. Statistics Canada, Health Analysis and Measurement Group, Minister of Industry, Ottawa, Canada, June. Catalogue 82-575-XIE. Downloaded from http://www.statcan.ca/english/freepub/ 82-575-XIE/2003001/pdf/report.pdf (accessed 16 May 2008).

101 Rachlis, M. (2004) *Prescription for Excellence: How Innovation is Saving Canada's Health Care System*, Harper-Collins Publishers Ltd, Toronto, Canada.

102 Leibowitz, R., Day, S. and Dunt, D. (2003) A systematic review of the effect of

different models of after-hours primary medical care services on clinical outcome, medical workload, and patient and GP satisfaction. *Family Practice*, **20** (3), 311–317.

103 Health, Canada (2004) A 10-year plan to strengthen health care. Downloaded from http://www.hc-sc.gc.ca/hcs-sss/delivery-prestation/fptcollab/2004-fmm-rpm/ index_e.html (accessed 16 May 2008).

104 Canadian Institute for Health Information (2006) Waiting for health care in Canada: what we know and what we don't know. CIHI, Ottawa, Canada. Downloaded from http://secure.cihi.ca/ cihiweb/products/WaitTimesReport_06_ e.pdf (accessed 16 May 2008).

105 Caplan, E. (2005) Realizing the potential of home care: competing for excellence by rewarding results. CCAC Procurement Review. Ontario Community Care Access Centres, Toronto, Canada. Downloaded from http://www.health.gov.on.ca/ english/public/pub/ministry_reports/ ccac_05/ccac_05.pdf (accessed 16 May 2008).

106 Randall, G.E. and Williams, A.P. (2006) Exploring limits to market-based reform: managed competition and rehabilitation home care services in Ontario. *Social Science & Medicine*, **62** (7), 1594–1604.

107 Williams, A.P., Barnsley, J., Leggat, S., Deber, R. and Baranek, P. (1999) Long term care goes to market: managed competition and Ontario's reform of community-based services. *Canadian Journal on Aging*, **18** (2), 125–151.

108 Baranek, P.M., Deber, R. and Williams, A.P. (2004) *Almost Home: Reforming Home and Community Care in Ontario*, University of Toronto Press, Toronto, Canada.

109 Williams, A.P. (2007) Strategic purchasing in home and community care across Canada: Coming to grips with "what" to purchase. *Healthcare Papers*, **8** (Sp), 93–103.

110 Kirby, M.J.L. and Keon, W. (2004) Why competition is essential in the delivery of publicly funded health care services. *Policy Matters*, **5** (9), 1–32.

111 Institute of Medicine of the National Academies (2006) Rewarding provider performance: aligning incentives in Medicare. Report Brief, Institute of Medicine, Washington, DC, September. Downloaded from http://www.iom.edu/object.File/master/37/236/11723_report_brief.pdf (accessed 16 May 2008).

112 Pink, G.H., Brown, A.D., Studer, M.L., Reiter, K.L. and Leatt, P. (2006) Pay-for-performance in publicly financed healthcare: some international experience and considerations for Canada. *Healthcare Papers*, **6** (4), 8–26.

113 Karha, J. and Topol, E.J. (2004) The sad story of Vioxx, and what we should learn from it. *Cleveland Clinic Journal of Medicine*, **71** (12), 933–939.

114 Mangano, D.T., Tudor, I.C. and Dietzel, C. (2006) The risk associated with aprotinin in cardiac surgery. *New England Journal of Medicine*, **354** (4), 353–365.

115 Hailey, D.M. (2007) Health technology assessment in Canada: diversity and evolution. *The Medical Journal of Australia*, **187** (5), 286–288.

116 Laupacis, A. (2002) Inclusion of drugs in provincial drug benefit programs: who is making these decisions, and are they the right ones? *Canadian Medical Association Journal*, **166** (1), 44–47.

117 Laupacis, A. (2005) Incorporating economic evaluations into decision-making: the Ontario experience. *Medical Care*, **43** (7), II-15–II-19.

118 Morgan, S., McMahon, M. and Mitton, C. (2006) Centralising drug review to improve coverage decisions: economics lessons from (and for) Canada. *Applied Health Economics and Health Policy*, **5** (2), 67–73.

119 Morgan, S.G., McMahon, M., Mitton, C., Roughhead, E., Kirk, R., Kanavos, P. *et al.* (2006) Centralized drug review processes in Australia, Canada, New Zealand, and

the United Kingdom. *Health Affairs*, **25** (2), 337–347.

120 Tierney, M. and Manns, B. (2008) Optimizing the use of prescription drugs in Canada through the common drug review. *Canadian Medical Association Journal*, **178** (4), 432–435.

121 McMahon, M., Morgan, S. and Mitton, C. (2006) The common drug review: a NICE start for Canada? *Health Policy*, **77** (3), 339–351.

122 Levin, L., Goeree, R., Sikich, N., Jorgensen, B., Brouwers, M.C., Easty, T. *et al.* (2007) Establishing a comprehensive continuum from an evidentiary base to policy development for health technologies: the Ontario experience. *International Journal of Technology Assessment in Health Care*, **23** (3), 299–309.

123 Borowski, H.Z., Brehaut, J. and Hailey, D. (2007) Linking evidence from health technology assessments to policy and decision making: the Alberta model. *International Journal of Technology Assessment in Health Care*, **23** (2), 155–161.

124 Oliver, A., Mossialos, E. and Robinson, R. (2004) Health technology assessment and its influence on health-care priority setting. *International Journal of Technology Assessment in Health Care*, **20** (1), 1–10.

125 Lalonde, M. (1974) A new perspective on the health of canadians: a working document. Minister of Supply and Services Canada, Ottawa, Canada. Downloaded from http://www.hc-sc.gc.ca/hcs-sss/alt_formats/hpb-dgps/pdf/pubs/1974-lalonde/lalonde-eng.pdf

126 Evans, R.G., Barer, M.L. and Marmor, T.R. (1994) *Why are Some People Healthy and Others Not: the Determinants of Health of Populations*, Walter de Gruyter Inc., New York.

127 Di Ruggiero, E., Frank, J. and Moloughney, B. (2004) Strengthening Canada's public health system now. *Canadian Journal of Public Health*, **95** (1), 5.

128 Basrur, S.V. (2005) Building the foundation of a strong public health

system for Ontarians. Annual report of the Chief Medical Officer of Health to the Ontario Legislative Assembly. Downloaded from http:// www.health.gov.on.ca/english/public/ pub/ministry_reports/ph_ontario/ ph_ontario_011706.pdf

129 Canadian Institute for Health Information, Statistics Canada (2007) Health Indicators, Canadian Institute for Health Information, Ottawa, Canada. Downloaded from http://secure.cihi.ca/ cihiweb/products/ hi07_health_indicators_2007_e.pdf

130 Evans, R.G. (2006) Fat zombies, pleistocene tastes, autophilia and the "obesity epidemic". *Healthcare Policy*, 2 (2), 18–26.

131 Canada. Parliament. Standing Senate Committee on Social Affairs, Science and Technology (2001) The health of Canadians: the federal role: volume one: the story so far. Interim report on the state of the health care system in Canada.

Ottawa, Canada. Downloaded from http://www.parl.gc.ca/37/1/parlbus/ commbus/senate/com-e/soci-e/rep-e/ repintmar01-e.htm (accessed 16 May 2008).

132 Canada. Parliament. Standing Senate Committee on Social Affairs, Science and Technology (2002) The health of Canadians: the federal role: volume two: current trends and future challenges. Interim report on the state of the health care system in Canada, January. Downloaded from http://www. parl.gc.ca/37/1/parlbus/commbus/ senate/com-e/soci-e/rep-e/ repjan01vol2-e.htm

133 Colombo, F. and Tapay, N. (2004) Private health insurance in OECD countries: the benefits and costs for individuals and health systems. OECD Health Working Papers, No. 15, Paris, France. Downloaded from http://www.oecd.org/ dataoecd/34/56/33698043.pdf (accessed 16 May 2008).

3
England

Adam Oliver

Population (million)	50.8
GDP per capita (US$ PPP)	38 000
Health spending as % of GDP	8.3
Public health spending as % of total spending	87.1
Health spending per capita (US$ PPP)	2724
Acute care beds per 1000 population	3.1
Practicing physicians per 1000 population	2.4
Life expectancy at birth (years)	79
Infant mortality per 1000 live births	5.1

Strategies used

1. Introduction of internal market
2. Establishment of National Institute for Health and Clinical Excellence
3. Hospital performance management
4. Patient choice of hospital provider

3.1
Introduction

Health care in England is predominantly publicly financed.[1] Specifically, the public sector accounts for well in excess of 80% of total health care expenditure, and the English National Health Service (NHS), which dominates health care provision, is more than 90% publicly financed. User charges are not therefore applied extensively in the NHS, and are only used to any noticeable degree in the provision of dentistry,

---

1) The health systems of the other countries that comprise the United Kingdom (i.e. Scotland, Wales and Northern Ireland) are broadly similar to the English system, but differ in some of their details. Unless otherwise stated, the English case will be the focus of attention in this chapter.

Cost Containment and Efficiency in National Health Systems: A Global Comparison
Edited by John Rapoport, Philip Jacobs, and Egon Jonsson
Copyright © 2009 WILEY-VCH Verlag GmbH & Co. KGaA, Weinheim
ISBN: 978-3-527-32110-0

optical services and pharmaceuticals, but even for these services exemptions are extensive. In the 2006–2007 financial year, the NHS had an expenditure of £84.3 billion [1], larger than the Gross Domestic Product (GDP) of Romania [2]. There is much concern in academic and policy circles that this expenditure be used 'efficiently', in terms of the amount of activity undertaken within the NHS, the quality/health outcomes it generates, or both. However, the concern with efficiency is not a recent policy invention: it has been central to the health policy debate for at least 20 years and usurped prior concerns of mere cost-containment. This chapter will review – and to some extent assess – the main efficiency-seeking initiatives that have been implemented in the NHS since the mid 1980s, including those that underlie the policy path pursued by the current Labour Government. First, however, a brief history of the structure and structural changes made to the NHS since the mid 1980s will be given, to enable the reader to place the efficiency-seeking attempts in their broader context.

3.2
The Structural Characteristics of the NHS

During the mid 1980s, Margaret Thatcher occupied 10 Downing Street, and the system prevailing in the NHS was – and had always been – widely cited as one of 'command and control'. Primary care doctors and dentists were self-employed but predominantly contracted to the NHS, while most other employees within the system were salaried and most hospitals were owned and managed by the state.[2] General practitioners (GPs) acted, and still act today, as gatekeepers to the use of nonemergency elective procedures. As now, the Department of Health (then called the Department of Health and Social Security) was allocated funds from central government, and the Department in turn allocated budgets, weighted by demographic and mortality data, to 14 regional health authorities. Each regional health authority was responsible for the strategic management of health care services in a geographically defined area, and they were collectively supported in this by a total of 192 district health authorities [4].

In the mid 1980s, the Conservative Government tried to introduce 'new public management' into the NHS, acting on the recommendations of the *Griffiths Report* [5, 6]. This Report called for an NHS management body, independent of the Department of Health, and headed by a Chief Executive. Hospital managers were given the task of holding health service personnel responsible for the levels, types and quality of their activities and for their levels of resource use, but the incentives for frontline NHS staff to improve their performance substantially were unfortunately insufficient.

---

2) Hospital doctors could and can in addition receive distinction awards and can spend time in the fee-for-service private sector [3], which may arguably serve as an incentive for them to under-perform in the NHS.

During the late 1980s, the Government developed radical proposals for the NHS [7–9], provoked by a general perception that the NHS was in a state of financial crisis [5, 10, 11]. *Working for Patients* [7] proposed a so-called 'internal market', where purchasers would agree contracts with competing providers. It was thought that the competitive nature of the market would provide the necessary incentives for the providers to improve efficiency. The internal market was introduced in 1991, and the purchasers were of two types [1]:

- The district health authorities, which were allocated budgets in order to purchase hospital care [2].
- GP fundholders, who were GPs that volunteered to hold a budget to provide primary health care and purchase some hospital care.

By 1996, some 50% of GPs were fundholders [11], while during the early 1990s *The Patient's Charter* [8] set a waiting times target of two years, and *The Health of the Nation* [9] introduced targets to reduce mortality rates for cancer, heart disease, stroke, mental illness, HIV/AIDS and sexual illnesses, and accidents. Until the Labour Party, under Tony Blair, was elected to government in 1997, the principal health care policy 'development' throughout the 1990s – other than the abolition of the regional health authorities – was to consolidate the reforms of the early 1990s.

In the years since the election of the Labour Government, health care has been an important policy concern. Perhaps most strikingly, prompted by a NHS funding 'crisis' in 1998–1999 [12], the Blair administrations have since 2000 committed themselves to making unprecedented increases in NHS spending so as to reach the average levels of expenditure in 'comparable' European countries by 2008 [13, 14]. As a result, the total expenditure on health care as a percentage of GDP at constant prices increased from 7.3% in 2000 to a projected 9.4% by 2008 [13, 14].

In 1998, a short time before initiating these increased spending commitments, the first Blair administration (1997–2001) published *The New NHS* [15]. This document proposed the replacement of the 'two-tiered' GP fundholding/nonfundholding system with 303 Primary Care Trusts (PCTs), which provide primary care and commission most secondary care. PCTs are financed by weighted capitation, and became fully operational in April 2004. They comprise GPs located in a particular area, supported by nurses, midwives, health visitors, social services and other stakeholders, and are run by managers. Moreover, during the late 1990s the district health authorities were replaced by 99 health authorities, which were later merged into 28 Strategic Health Authorities (SHAs). Since PCTs are now the principal purchasers of secondary care, other than retaining commissioning responsibilities for highly specialized health services, the role of the SHAs is merely one of monitoring the performance of PCTs and hospitals.

Also during the late 1990s, the National Institute for Clinical Excellence (NICE) and the Commission for Health Improvement (CHI) were established to pursue better quality, efficiency and consistency within the NHS. NICE has a remit to assess new and existing interventions for their clinical and cost-effectiveness, and to decide whether an assessed intervention ought to be made available in the NHS. CHI (now called the Healthcare Commission) is an inspectorate set up to monitor NHS quality, performance and adherence to NICE guidance. In short, NICE sets the standards, and CHI was set up to monitor these standards.

The Labour administrations embraced two other health policy areas that may appear contradictory: a concern with health inequalities, and a greater involvement of the private sector. With respect to health inequalities, the first Blair administration commissioned the *Acheson Inquiry* in 1997 [16], made the narrowing of the health gap an explicit aim in the consultation document *Our Healthier Nation* [17], and issued a set of social class and geographic-related health inequalities targets to be met by 2010 [18]. Regarding the private sector, much capital investment in NHS Trusts was directed through the private finance initiative (PFI), where private firms are contracted to build facilities and operate nonclinical ancillary services [19]. The PFI has attracted criticism amid claims of profiteering by private consortia [4], and concerns that private sector borrowing may ultimately prove more costly than public sector borrowing [20–23].

A second notable development in the closer integration of the private sector is the private sector concordat [24, 25], which allows the purchasers of health care to commission private-sector facilities in order to reduce waiting times for elective surgery. If necessary, private providers located away from the area in which the patient resides, and even overseas providers, can be used. Use of the private sector in this way is linked to the Labour Government's movements to extend patient choice, which in January 2006 required GPs for the first time to offer patients requiring elective surgery a choice of initially four or five hospital providers at the point of referral [26], but with the intension of widening this choice to all available public and private providers by 2008.

Underlying the choice plans is a system of hospital payments introduced in September 2004, termed Healthcare Resource Groups (HRGs) [27]. Hospitals are offered a set, national price per procedure defined by the HRG system, and therefore will not be able to compete for patients on the basis of price. The logic thus appears to be one of encouraging competition on the basis of quality. By the 2007–2008 fiscal year, the expectation is that 90% of hospital expenditure will be covered by these national tariffs [28, 29].[3)]

Overall, the main emphasis in government health care policy initiatives over the past two decades has been on attempting to improve supply side efficiency, probably primarily as a means to reduce waiting times. Figure 3.1 provides a simple diagrammatic depiction of the NHS as it now stands. The majority of the remainder of this chapter will focus in a little more detail on these efficiency initiatives and, where possible, will attempt to detail the effects that they have had.

---

3) The 90% aspiration may be a little optimistic, in that the national tariff was supposed to apply to 80% of hospital activity by 2005–2006, but in the event applied only to elective care, substantially less than 80% of all activity.

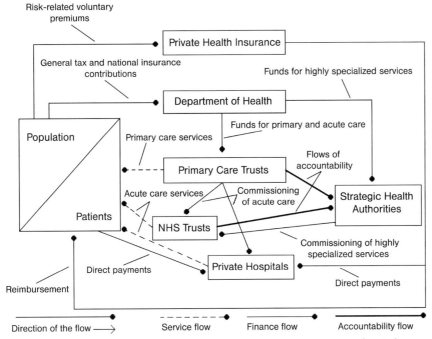

Risk-related voluntary premiums

Private Health Insurance

General tax and national insurance contributions

Funds for highly specialized services

Department of Health

Population

Primary care services

Funds for primary and acute care

Flows of accountability

Primary Care Trusts

Acute care services

Commissioning of acute care

Strategic Health Authorities

Patients

NHS Trusts

Commissioning of highly specialized services

Direct payments

Private Hospitals

Direct payments

Reimbursement

Direction of the flow ⟶   Service flow   Finance flow   Accountability flow

Figure 3.1 Structure of the English health care service. Note: NHS Trusts are NHS hospitals.

3.3
The Efficiency Initiatives

3.3.1
The Internal Market

Up until very recent years, the internal market experiment was the most radical reform initiative in the history of the NHS, and was focused primarily on increasing productive efficiency (i.e. increasing activity per unit of cost). Prior to the introduction of the internal market, hospitals were financed mainly through weighted capitation from the health authorities. The internal market introduced a 'purchaser–provider split', whereby it was intended that the hospitals compete for contracts from the GP fundholders and the district health authorities to provide health care services. GP fundholders were allowed to retain budget surpluses, provided that these were reinvested into their practices, potentially increasing the value of the GP's saleable assets. When fundholding was first introduced, there was some concern that making GPs financially responsible for referrals to secondary care would result in risk selecting, or 'cream-skimming', which would have possible implications with respect to equity of access. However, no evidence of cream skimming was observed [5]. Overall, there was a belief by government that the internal market reform would offer the necessary incentives to hospitals to compete on the basis of price, quality and activity levels, stimulating improvements in efficiency.

It is not easy to assess the 'whole system' impact of health care policy initiatives, not least because there are so many confounding factors. However, up until the early 2000s the Department of Health did produce (albeit imperfect) indexes for hospital activity and expenditure, which made possible some estimate of trends in NHS productive efficiency over time, and with which an insight into the impact of the internal market can be gleaned. The activity index combined some selected hospital and community health services activities into a single number, while the expenditures index served as an indicator of resource use in those services. Dividing the activities index by the expenditures index gives an index for activity per unit of resource use, which is an indicator of productive efficiency. These indexes for the period 1985–2001 are presented in Table 3.1.

In Table 3.1, the indexes are fixed at 100 for the 1974–1975 financial year. This means that activity was 27.9% higher in 1985–1986 than it was in 1974–1975. The figures in parentheses show the percentage year-on-year increases. The internal market was introduced in 1991, and thus the data in the table indicate that the most sustained period of relatively large increases in activity occurred between the advent of the internal market and 1996. Therefore, the internal market may initially have had an impact on activity, although the accompanying substantial increases in health care expenditure during the early 1990s presents an attribution problem, in that the increased activity may have been due to the extra expenditure, rather

Table 3.1 Trends in activity per unit of cost; 1974–75 = 100.

Year	Activity index	Expenditure index	Activity per unit of cost
1985–1986	127.90 (2.67)	112.00 (0.19)	114.19 (2.5)
1986–1987	129.80 (1.49)	112.36 (0.32)	115.52 (1.2)
1987–1988	131.92 (1.63)	113.22 (0.76)	116.52 (0.9)
1988–1989	133.05 (0.86)	114.03 (0.72)	116.68 (0.1)
1989–1990	135.98 (2.20)	116.02 (1.74)	117.21 (0.5)
1990–1991	137.75 (1.30)	117.11 (0.94)	117.63 (0.4)
1991–1992	144.95 (5.23)	120.16 (2.60)	120.64 (2.6)
1992–1993	149.45 (3.10)	123.89 (3.10)	120.63 (0.0)
1993–1994	155.41 (3.99)	125.86 (1.59)	123.48 (2.4)
1994–1995	161.91 (4.18)	127.61 (1.39)	126.88 (2.8)
1995–1996	168.30 (3.95)	129.85 (1.76)	129.61 (2.2)
1996–1997	171.10 (1.66)	131.76 (1.47)	129.85 (0.2)
1997–1998	174.18 (1.80)	134.68 (2.21)	129.33 (−0.4)
1998–1999	177.77 (2.06)	138.65 (2.95)	128.21 (−0.9)
1999–2000	179.69 (1.08)	143.81 (3.72)	124.95 (−2.5)
2000–2001	179.69 (0.00)	150.34 (4.54)	119.53 (−4.3)

Source: Department of Health, personal communication.

than the market *per se* [4]. By the late 1990s, the relatively high annual activity increases had ceased.[4]

The expenditure index shows relatively high annual increases in the early 1990s, as alluded to above, and in the late 1990s/early 2000s; this was consistent with the large increases in NHS expenditure in these two periods. Although the internal market did appear to spark some improvement in productive efficiency, as measured by activity per unit of cost, the large increases in expenditure from the turn of the century, coupled with no activity increase, resulted in a decline in productive efficiency at that time. One could of course retort that since it takes several years to train doctors and nurses, the impact of the increases in NHS expenditure will have a lagged effect, with activity increases becoming apparent several years downstream. However, it is worth noting that the European Working Time Directive [31], which limited/will limit the working week of physicians to 58 h in 2004 and 49 h in 2009, will in itself absorb the equivalent of 12 550 doctors [32].

On the basis of the above evidence, a case can be made that the internal market reform of the early 1990s had only a short-term effect on productive efficiency, moving the system to a slightly higher baseline. It may be contended that the attainment of a higher 'plateau' was worth the effort, although even some of the early proponents of the internal market have questioned whether its benefits have exceeded its costs [33]. If one does draw the conclusion that the internal market was not a resounding success, then several plausible explanations can be given:

1. Since hospital doctors are paid primarily by salary, they had no direct incentive to increase their activity levels substantially, although they may have faced some pressure from hospital managers, eager to secure contracts from the purchasers of health care.
2. GP fundholders were possibly too small to bargain effectively in contract negotiations with the often large, powerful hospital providers, and may have lacked adequate information sources to enable them to make informed decisions [34].
3. It may have been the case that, after only a few years, the hospitals had exhausted their capacity to move to shorter hospital stays and day-case surgery, which would inevitably have put a halt on increased throughput [35].
4. It is possible that the introduction of the internal market had the initial psychological effect of inducing people to alter their behaviors, but once the novelty of the policy had dissipated a new equilibrium was reached which did not motivate the acceptance of increased workloads. The market may have also provoked relatively short-term changes in behavior on the purchaser side of the contract, with respect to more frugal prescribing [36] and referral [37] patterns by GP fundholders.

---

4) Martin and Smith [30] have written that long-term increases in expenditure are required to increase activity and reduce waiting times, and that a short-run boost to surgical capacity will ultimately fail to bring about the desired effect. Later in this chapter it will be shown that significant reductions in waiting times and long-term increases in health care expenditure after 2000 were correlated events.

5. The organizational institutions of the NHS are not designed to facilitate competition [38, 39]. For example, many GPs and health authority managers had longstanding relationships with the doctors working in their local hospitals, and would have in many cases been unwilling to undermine the financial position of the hospitals in which these specialists work. In addition to these collegial networks, the NHS is characterized by hierarchical relationships, with the government ultimately held accountable by the electorate for perceived cutbacks to the system. Politically therefore it is very difficult for any government to allow poorly performing hospitals to close [13], as was illustrated in the 2001 and 2005 general elections when a physician was elected and re-elected to parliament on the single issue of opposing the closure of his local hospital. Alain Enthoven, perceived by many to be the intellectual architect of the internal market [40], has indirectly concurred with this view, citing the Government's reluctance to let the full force of market competition flow as the reason behind the internal market's limited impact [41].

6. By 1996 it was clear to most people that the Labour Party was in a very strong position to be elected to government in 1997, and that they had committed themselves to abolishing the internal market. Once elected, the Labour Government actually retained – indeed, universalized – the internal market, with the PCTs acting as the purchasers of health care, but the expectation that the market would end in 1997 may have undermined the various actors' motivations to pay attention to the incentives of the mechanism.

Before leaving the subject of the internal market, it is noteworthy that the Government decided to no longer construct the indexes listed in Table 3.1 beyond the 2000–2001 financial year. A cynic might conclude that the Government took this decision because the data did not place their health policy program and increased health care expenditures in a positive light, but there were also good, scientific reasons for terminating these measurements. For example, the activity index included a very small number of services that were disproportionately biased towards inpatients. Moreover, it gave no credit for a complex case-mix, it could be increased by providing more of the most expensive and potentially cost-ineffective care, and it ignored quality, patients' experiences and health outcomes, all of which could deteriorate with increased activity. Alternative methods that address at least some of these problems are being developed to measure NHS productivity and expenditure [43], but as yet, no definitive measure of productivity is available [42].[5)]

---

5) The new methods have focused upon only the post-1995 period, but most have shown, at best, a 4% decline in productivity over the 1997–2003 period [41]. An exception is a recent study by Martin et al. [42], who incorporate patient survival rates, health status gains and adjustments for the life expectancy distribution of patients and the stress-negating effects of reduced waiting times in their measurements, and report an annual average increase in productivity of 0.17% over the 1999–2000 to 2003–2004 period.

3.3.2
The National Institute for Health and Clinical Excellence[6]

The internal market experiment was primarily focused upon improving efficiency with respect to activity. During the late 1990s, the Labour Government established NICE, which had an initial remit to end geographic variations in prescribing patterns. NICE can be seen as the culmination of broader developments in health technology assessment (HTA) in the 1990s which, according to Stevens and Milne [45], gained momentum over that period due to the introduction of the internal market, with the argument being that the internal market incentivized purchasers of health care to demand better 'value for money'. NICE, however, is less concerned with activity efficiency; rather, its focus is on health outcomes efficiency.

NICE assesses the clinical and cost-effectiveness of a number of selected health care products and services, recommending the use of cost-utility analysis (CUA) and thus the employment of quality-adjusted life years (QALYs) as the preferred health outcome measure. The underlying assumption of CUA, and thus of NICE, is that the relevant goal in the allocation of health care resources is the maximization of health inside the budget constraint, a health-specific utilitarian approach. NICE is considered by many as the most sophisticated HTA agency in the world, in that it combines both scientific assessment and policy appraisal into a single decision-making body [45], and it has also been lauded for its transparency [46].

NICE guidance to the NHS was made mandatory in 2001, and PCTs must not deny funds for treatments that have been recommended for use by NICE when more than three months have passed since the guidance was issued [45]. Unfortunately, this may steer the NHS towards a suboptimal focus upon those interventions that NICE assesses [45, 47]. The threshold at which NICE deems interventions to be cost-effective, although implicit, appears to be somewhere between £20 000 and £30 000 per QALY gained [48], with the more generous end of this range being employed when an intervention has particular positive characteristics; for example, if the intervention is especially innovative [47]. This threshold has been criticized as being too generous, out of line with NHS products and services not assessed by NICE [49], while Williams [50] argued that the NHS should pay no more than the average per capita GDP of about £18 000 for each QALY gained. However, NICE faces huge media-driven public pressure when it rules against the provision of interventions, emphasized by recent decisions vis-à-vis drug therapies for multiple sclerosis (MS). Indeed, given the political difficulties of recommending against the use of treatments, NICE guidance often only goes so far as to advocate restricted use in certain patient categories [48], which can upset physicians, many of whom believe that a broad brush guidance – even when applied in relation to specific categories – tends to

---

6) NICE was originally called the National Institute for Clinical Excellence, but on having its remit extended to include public health interventions (partly in response to criticism that its original remit did not include interventions that probably have a greater impact than clinical services on population health [44]), it is now called the National Institute for Health and Clinical Excellence.

overlook the very 'special' situation of many patients, including the presence of comorbidities, particular socioeconomic circumstances, and so on [51]. NICE seems almost destined to be damned for going 'too high' or 'too low' in its choice of a cost-effectiveness threshold, depending on the perspective of the particular critic.[7] Further criticisms that could be waged against the implicit threshold are that it lacks a scientific justification [45, 49], and that the threshold should be explicit rather than implicit. A rejoinder could be made, however, in that if the threshold was explicit, pharmaceutical companies would have the direct incentive to price their products at marginally below the threshold, irrespective of whether they could have in actuality priced much lower.

A serious methodological problem exists in the framework of CUA (and broader forms of cost-effectiveness analyses) itself [47, 52], which NICE has yet to adequately address, even though the problem has been noted explicitly by managers and health economists who sit on the NICE advisory committee [53]. The problem is that 'cost-effective' interventions are usually 'cost-increasing', which leaves PCT managers unsure as to where to find the extra required resources, a situation that is likely to worsen with the termination of the substantial increase in NHS expenditures witnessed in recent years. NICE provides no guidance as to where the required resources are to come from, which places the PCT managers in the position of having to cut back on existing, nonassessed interventions, if it is politically feasible for them to do so. Unfortunately, these nonassessed interventions may in fact represent better value for money than those that replace them. Appleby *et al.* [49] suggest that the nonassessed products and services may indeed be more cost-effective than those recommended by NICE, which is why they have called for a lower cost-effectiveness threshold; however, we then re-enter the political difficulties of NICE recommending against too many interventions. In NICE's defense, it has no remit over the affordability of health care; unfortunately, it is not possible to divorce considerations of efficiency from those of affordability [52]. The problem outlined above is not, however, intrinsically with NICE, but with the form of economic evaluation that it currently recommends.

Further concerns have been cited. For instance, some have written that NICE has insufficient capacity to assess a meaningful number of interventions, and/or that it has traditionally focused too heavily on new interventions, overlooking to a substantial extent the potentially large number of cost-ineffective interventions in the NHS [54]. In a recent study, Linden *et al.* [55] examined 159 technologies reviewed from 88 appraisals between March 2000 and June 2006, and found that 84 (53%) were new technologies and 75 (47%) were existing technologies; on this basis the authors argued that this did not indicate a bias towards new technologies. It is worth bearing in mind, though, that the number of existing technologies far outnumbers the number of new interventions, and therefore an approximately 50 : 50 split may still

---

7) Those motivated by technical arguments may argue that NICE is, at present, too generous, but those motivated by political factors are more likely to argue the opposite.

represent a significant bias towards assessing the newer products and services. The limited capacity of NICE may create legitimate cause for concern that its impact will inevitably be limited, although one could counter this concern by arguing that the producers of medical interventions face the risk that their products *may* be assessed and thus have the incentive to make their products more cost-effective than they would be otherwise.

Despite the transparency of NICE, with its evidence and value judgments published in detail and available for expert and – in theory – public scrutiny, the level of technical sophistication now applied in measurements of costs and outcomes somewhat undermines any claim that HTA-based decision making affords greater transparency to the public (and maybe even to most experts) [56]. This view has also been expressed by many members of the NICE appraisals committee [48, 53]. It is particularly worrying when one bears in mind that the methods of economic evaluation are far from perfect (and may even do more harm than good).

There is also a small research industry assessing the appropriateness (or otherwise) of the health maximization assumption that underlies standard forms of economic evaluation [57]. The concern is that health maximization – an efficiency argument – fails to account for equity considerations in the distribution of health care resources, although most of the work in this area has focused upon equity in health outcomes rather than equity of access. The very tentative finding in the literature is that many members of the general public would trade-off some total health for a more equal distribution of health across particular groups, such as those defined by income and/or social class. Even if this is the case, it remains debatable as to whether the NHS should serve as the vessel through which to attempt to redistribute health outcomes, not least because this would entail the systematic prioritization of those in lower social classes over those in higher social classes, or indeed the employment of different cost-effectiveness thresholds for different social/income groups [47]. This is a potentially counterproductive policy direction when one remembers that the NHS relies on middle class tax contributions for much of its revenue, even without recourse to the ethical concerns around systematically prioritizing some people over others for life-saving operations based on the size of their income. It may be wiser to attempt to address any socially unacceptable inequalities in health through wider social and fiscal policy. Whatever one's opinion on this matter, NICE has yet to incorporate equity concerns systematically into its assessment process, and it is in any case possibly more pertinent for policy makers to focus upon equity of access rather than equity in health outcomes in the NHS. More will be written on this subject in the conclusion to this chapter.

Chinitz [58] has argued that the highly centralized structure of the NHS has facilitated a relatively easy introduction of national guidance and guidelines compared to other countries. However, the actual policy impact of NICE guidance on care quality and variations has been assessed as 'variable' [59]. Although one ought to acknowledge that things may change with time, it seems as though NICE guidance has had the largest effect when it has run parallel to the support of opinion leaders, professional bodies and marketing efforts by pharmaceutical companies [59]. Indeed, some have even pondered whether NICE in itself is cost-effective [60].

NICE can, however, be seen in a more positive light. Many of the problems discussed above are methodological, and it may be possible to address these adequately with time. Moreover, one must consider NICE with respect to the pre-NICE era, where GP and hospital prescribing was influenced heavily by pharmaceutical companies and powerful hospital consultants and interest groups [45, 46]. The political path of least resistance is to say 'yes' to powerful lobbies and 'no' to weaker lobbies, irrespective of the products and services on offer (Richard Cookson, personal communication). Moreover, economic evaluation may work sufficiently well when interventions are of high cost but offer little benefit, which may not be such a rare occurrence in any health care system. Overall, although NICE is beset by problems, its underlying motivation is honorable, and although the jury is still out it may, with time and with the appropriate developments, positively benefit the populations of England and Wales.

3.3.3
Performance Management

NICE can be perceived as an agency of performance management, whereby local purchasers and providers of health care are held accountable to act upon centrally issued guidance. The love affair with performance management under the first and second Blair administrations was not, however, limited to NICE. In 2001, the Government introduced a hospital 'star rating system', whereby NHS hospitals were assessed annually on a number of indicators, including targets concerning waiting times, cleanliness, treatment-specific data and financial management. CHI assumed the responsibility for monitoring the hospitals' performance from the Department of Health in 2003 and, at the extremes, hospital management teams could be either dismissed or earn greater autonomy on the basis of CHI's assessments [61]. Although the star rating system was terminated in 2005, hospitals are still subject to inspection in a similar way through the 'health check' undertaken by the Healthcare Commission (http://annualhealthcheckratings.healthcarecommission.org.uk/annualhealthcheckratings.cfm) and, on the back of good performance, can gain 'Foundation Trust' status. This allows the hospital to retain revenues from land sales, to determine its own investment plans, and also offers scope for it to offer additional performance-related rewards to its staff [13, 62, 63].

Bevan and Robinson [13] have cited some evidence that, they argue, lends support to the star rating system. First, the national target for ambulances to reach 75% of life-threatening emergencies within 8 min, as introduced in 1996, became a key target of the hospital rating system in 2002–2003, and dramatic improvements in hospital performance against this target were noted as a consequence.[8] Second, a key target in the star rating system in England in 2002–2003 was that no patient should wait more

---

8) In 2000, the ambulances at some hospitals were reaching only 40% of these cases within 8 min, but by 2003, the worst-performing hospital was reaching almost 70% [13].

than 12 months for elective acute care, and by March 2003 this target had been met almost entirely. On the other hand, in Wales, where there was no star rating system, 16% of patients were still waiting for more than 12 months.

Waiting times targets had long preceded the star rating system, and thus the success in reducing waits in England cannot be attributed entirely to star rating *per se* (although the incentives to meet the targets have perhaps played an important role over recent years). Indeed, Hauck and Street [64], albeit by examining data from only four hospitals, have noted that performance against waiting times became better in England as compared to Wales after the latter abandoned waiting times targets in 1998. When Wales reintroduced such targets in 2001 (though not as part of a star rating system) there was some convergence in waits across the two countries.

Incidentally, for the media and the general public, waiting times are the key indicator of NHS performance, probably because they are easy to understand and were, until recent years, unacceptably long. Consequently, there has been an almost constant call for their reduction. Waiting time targets have therefore become a political imperative,[9] and were a focus within *The Patient's Charter* in the early 1990s [8], which guaranteed admission to treatment within two years. The percentage of patients who waited more than two years for treatment fell from 10% to 0% between 1989 and 1993 [11], and although the percentage was declining prior to 1991, *The Patient's Charter* may have contributed towards the overall trend. *The Patient's Charter* was extended in 1995 to guarantee a maximum waiting time of 18 months. Now, no-one waits more than 18 months for ordinary or day-case admission, and it therefore appears that the waiting time targets established by the Conservative administrations during the 1990s have largely been achieved.

The Labour Government, through *The NHS Plan* [24] and the *Wanless Review* [66], has proposed a number of waiting time targets, as outlined in Table 3.2. The recent trends in waiting times for ordinary and day-case admissions combined are presented in Table 3.3, where the data show that by March 2007 the six-month waiting

Table 3.2 Waiting time targets taken from *The NHS Plan* and the *Wanless Review.*

Year	Maximum inpatient waiting time
2002–2003	15 months
2005–2006	6 months
2008–2009	3 months
2022–2023	2 weeks

Sources: Department of Health [24], Wanless [66], Yates [65].

9) There is a distinction between waiting lists and waiting times, even though there is an association between the two measures [65]. It is generally thought that waiting times are a more sensible target in that waiting list reductions can be met by focusing health care resources towards those with minor ailments who have been waiting for relatively short periods of time. Therefore, the issues around waiting times will be considered in this chapter.

Table 3.3 Trends in waiting lists and waiting times for ordinary
and day-case admissions combined in England.

Year	Months waiting (% of total)		
	<3	<6	<12
March 1999	51	74	96
March 2001	52	76	96
March 2003	50	76	98
March 2005	70	95	100
March 2007	85	100	100

Source: Department of Health [67].

time target had been met, and a relatively small percentage of people continued to wait more than the 2008–2009 target of three months to receive treatment.

The downward trend in the waiting times figures are probably predominantly due to a combination of targets, incentives to meet those targets, and the increases in NHS expenditures. There are of course many possible problems with targets; for example, the incentive for hospitals to manipulate statistics, the potential for targets to distort priorities at the expense of other desirable but untargeted objectives, and the difficulty in isolating the specific effects of targets due to gaming and confounding factors [64]. However, on balance it does appear that the Government has used targets effectively to achieve some success in reducing waiting times.

Away from the hospital sector, the Government has also used performance management in the primary care sector. In April 2004, the Government introduced the new GP contract, with 18% of GP income depending on their performance against 146 indicators of clinical quality, practice organization and patient experience [68]. The definitive, overall success of the GP contract remains difficult to gage; GPs have seemingly performed well against the targets – in fact too well, in that the Government's underestimation of how far GPs would comply with the performance indicators contributed towards some financial overspend concerns in the NHS in 2006, unexpected in the era of unprecedented increases in NHS expenditure. Moreover, the GP contract may suffer from the general problems of targets cited above, including gaming by GPs and a possibly increased readiness to prescribe cost-ineffective care to meet particular clinical targets. Smith [34] has suggested that the use of indicators in the NHS is often opportunistic and selective, relying on existing data sources rather than a 'rational' selection of indicators that would genuinely improve the NHS (e.g. that good performance against process indicators improves population health), and therefore an ongoing assessment of the outcomes of GP performance against the contract is required. Moreover, care must be taken to ensure that performance management does not alienate the professionals on which the NHS relies. All things considered, however, performance management through the contract has indicated that GPs respond well to incentives, and it is possible that the contract has improved the NHS. Moreover, similar incentive mechanisms have seemingly worked well in other health care systems [69].

3.3.4
Patient Choice

In very recent years there has been a shift in emphasis in English health policy from targets to markets, brought about by the Government's dissatisfaction that their heavy expenditures in the NHS were not working sufficiently and their consequent consultation with a number of pro-market policy entrepreneurs.[10] Married to the notion of a market, the Government has embraced the notion of patient choice, specifically since January 2006, by requiring that GPs offer patients a choice of hospital at the point of referral. Indeed, on the basis of patient reports, the offering of a choice of hospital is now included in the GP contract. Facilitating the patient choice policy direction is the HRG payment system of national tariffs for hospital procedures. The idea is that since prices are fixed, hospitals will want to compete on the basis of quality for patients in order to receive the HRG payments with, in effect, the money following the patient. For example, if hospitals that have excess capacity offer relatively shorter waiting times, the choice may decrease any excess capacity in the whole system, increase activity, and reduce waiting times. Boyle [70] has, however, raised the concern that the HRG payment may give an incentive for high-quality providers to skimp on quality, in order to reduce costs.

It is too early to assess the impact of patient choice,[11] but a number of concerns can be raised. First, in a system were people are accustomed to their GP acting as their agent for care, it is plausible that many patients – particularly the elderly and uneducated, and those with moderate to serious health conditions – would want their GP to continue to choose for them (indeed, in order to make a 'rational' choice, hospital performance data may often not be understandable or sufficient for most patients, citizens or even GPs). Consequently, the collegial networks among doctors mentioned in Section 3.3.1 are likely to remain intact, and thus patients will be referred to the same hospitals to which they have been referred in the past.

Second, many people may use simple decision rules in formulating their choices, basing their decision on what is for them the most prominent attribute of the options before them. This has to some extent been observed with respect to the choice of hospital among those unfamiliar with such choices. Survey evidence has shown that people often focus upon how far they would need to travel to the various options [71, 72], and hence there may be a tendency for people to choose their local hospital for reasons of convenience to themselves and friends/family. Consequently, the choice may well have very little impact on where they are treated. Those in favor of more choice could retort by citing evidence that choice may work to reduce waits and cause waits to converge across hospitals in large urban areas (i.e. London) where

---

10) As noted in the previous section, however, one could quite plausibly argue that the target-based health policy direction was/is beginning to bear fruit.

11) One could of course attribute some of the trend in reduced waiting times summarized in Table 3.3 to the patient choice policy, but this trend was apparent before choice was extended and is perhaps more likely to be the result and the targets and increased expenditures discussed earlier.

there are many hospitals within reasonably close proximity to each other [73],[12] and/ or that, over time, people will increasingly act upon choice once they become accustomed to it being offered. Moreover, it may be the case that the uptake of choice has only to be small in order to create the necessary incentives for a substantial improvement in performance across most hospitals – an argument that mirrors that pertaining to NICE's assessment of only a few products motivating all companies to improve the value for money of their products.

However, concerns can be raised if people do indeed act upon the opportunities to choose their hospital. For instance, if a sufficient number of people embrace the notion of choice, and demand immediacy in their health care treatment from their own individual perspective, this may place an impossible strain on the collective finances of the NHS, particularly when the large increases in NHS expenditure come to a halt in 2008. The Government has recognized this potential problem, and thus to serve as an expenditure stop on the potentially escalating costs of choice, has introduced practice-based commissioning (which is essentially a reintroduction of GP fundholding) in order to incentivize responsible prescribing and referral patterns that remain within planned expenditure ranges.[13] The Government also hopes that this will incentivize more efficient practices by GPs, although whether this will work, given the continued existence of some of the constraints that undermined the first incarnation of the GP fundholding experiment, is a moot point. Nonetheless, a tension between extending patient choice and controlling costs is apparent, and will become even more apparent if HRGs result in price increases [29].

Although some hospitals may well improve their performance in the face of increased competition for patients, there are likely to be other hospitals that cannot legitimately improve, and indeed some hospitals that might deteriorate due to declining funds, yet patients will still have to use them. It is plausible that extending patient choice in the NHS will, at the societal level, do more harm than good, although it is perhaps important to point out that the debate on this issue has thus far been largely ideologically driven. It will be some time before the success or otherwise of this policy direction can be assessed empirically.

3.4
Conclusions

It is hopefully clear from the above discussions that efficiency concerns have dominated the NHS policy direction since the early 1990s, and the recent patient choice agenda brings us to the present. Some of the policy initiatives appear to have

12) In the London Patient Choice Project (LPCP), there was an increase in the total funding of surgical capacity in London [73]. It would be interesting to learn whether a similar success of extending choice is observed in a more resource-constrained environment, and since the LPCP reported success in oph-

thalmology and orthopedics only, across a wider range of specialties.

13) Prior to practice-based commissioning, budget holding responsibility was held by PCT managers. Now, the prescribers of care – that is, the GPs – hold indicative budgets.

been more successful than others, and for some it is too soon to ascertain definitively their impact. However, the success of these initiatives cannot sensibly be assessed in isolation of other important NHS objectives.

For example, according to policy documents, the core aims of the NHS are (and always have been) universality, comprehensiveness and largely free care at the point of use [4, 24] – aims that lend themselves to the principle of equal access for equal need. Universality refers to the whole population being eligible to use the system; comprehensiveness refers to the range of services on offer, and is something of a moving target because as a country becomes wealthier, the specified basket of services, and the quality of those services, may have to increase to ensure that comprehensiveness is maintained. The basket of health care services in the NHS has generally increased in line with international developments in health technology. Thus, on the whole, the principle of universal, comprehensive care that is largely free at the point of use has been protected. However, it is possible that the continued pursuit of efficiency through, for example, extending choice and attempted maximization of health outcomes might, without due caution, undermine this principle in the future.

For instance, consider patient choice. If, as discussed above, choice does provide cost pressures that GPs would find difficult to counter in the face of rising patient expectations, and if the intrinsic value of choice for patients and citizens renders it politically difficult for the Government to retract choice once introduced, then other core goals of the NHS – such as universality, comprehensiveness and/or affordable care at the point of use – may have to be retracted instead. If those who believe in choice also believe in the other NHS objectives, then they ought to explicitly acknowledge that there is a potential trade-off between objectives, and they should exercise caution and clarity over where choice is to be encouraged and exercised.

The founding aims of the NHS could also be compromised by the focus upon health outcomes in health economic evaluation. Even if the methods of economic evaluation did not suffer from the limitations discussed earlier, a great many examples could be given to illustrate this point, but for illustrative purposes just one will be given here. Assume that there is currently no available treatment for, say, MS, but that treatment is available for angina pectoris. Assume further that new treatments for each of these conditions are developed, but that the decision maker can fund only one of them. The decision maker is informed that the new angina intervention produces more QALYs per additional required unit of resource than the MS intervention, and that therefore, following the rules of CUA, the angina intervention ought to take priority. However, given that each of us could at some point in the future suffer from angina and/or MS, it is quite possible that most people would in these circumstances prefer the MS intervention to take priority, in order to provide the security of knowing that some form of public sector intervention will be provided if one were to suffer from either of these conditions. In this context, systematically prioritizing the angina patients may violate universality because the MS patients would still be without access to the health care system.

Notwithstanding the tensions between 'old' and 'new' health care policy objectives, a consumerist emphasis on competition and choice will probably be an important part of government health policy for the foreseeable future. Despite the theoretical concerns with the choice agenda, it may well achieve its aims; only time will tell. Moreover, health economic evaluation does show some promise in specific contexts, if only for providing a justification for the nonfinancing of costly, largely ineffective interventions. However, a myopic pursuit of efficiency gains without acknowledging or even recognizing their trade-off with older, perhaps even more important health care equity objectives could fundamentally undermine the collectivist nature of the NHS. This author is cautiously optimistic, however, that it will not come to this, and hopes that a balanced pursuit of all objectives without allowing any one aim to fundamentally undermine all others will shape the future of NHS policy.

References

1 Department of Health (2006) *Departmental Report*, Department of Health, London.

2 Hewitt, P. (2007) Speech by the Secretary of State for Health, London School of Economics: London (http://www.lse.ac.uk/collections/LSEHealth/eventsAndSeminars/Jointpubliclectures/LSE%20DRAFTSPEECH FINALASDELIVERED.doc) (14 June 2007).

3 Maynard, A. and Bloor, K. (2003) Do those who pay the piper call the tune? *Health Policy Matters*, 8, 1–8.

4 Webster, C. (2002) *The National Health Service. A Political History*, Oxford University Press, Oxford.

5 Klein, R. (2001) *The New Politics of the NHS*, 4th edn, Pearson Education, Harlow.

6 National Health Service Management Inquiry (1983) *Report*, DHSS, London.

7 Department of Health (1989) *Working for Patients*, Cm. 555, HMSO, London.

8 Department of Health (1991) *The Patient's Charter*, HMSO, London.

9 Department of Health (1992) *The Health of the Nation. A Strategy for Health in England*, Cm. 1986 HMSO, London.

10 Hacker, J. (2004) Dismantling the health care state? Political institutions, public policies and the comparative politics of health reform. *British Journal of Political Science*, 34, 693–724.

11 Le Grand, J. and Vizard, P. (1998) The National Health Service: crisis, change or continuity? in *The State of Welfare: The Economics of Social Spending*, 2nd edn, (eds H. Glennerster and J. Hills), Oxford University Press, Oxford, pp. 75–121.

12 European Observatory (1999) *Health Care Systems in Transition. United Kingdom*, WHO, Copenhagen.

13 Bevan, G. and Robinson, R. (2005) The interplay between economic and political logics: path dependency in health care in England. *Journal of Health Politics, Policy and Law*, 30, 53–78.

14 Department of Health (2002) *Delivering the NHS Plan*, Cm. 5503, Stationery Office, London.

15 Department of Health (1998) *The New NHS. Modern, Dependable*, Stationery Office, London.

16 Department of Health (1998) *Inequalities in Health. report of an independent inquiry chaired by Sir Donald Acheson*, Stationery Office, London.

17 Department of Health (1998) *Our Healthier Nation*, Stationery Office, London.

18 Department of Health (2002) *Tackling Health Inequalities: the Results of a Consultation Exercise*, Department of Health, London.

19 Le Grand, J. (2002) Further tales from the British National Health Service. *Health Affairs*, 21, 116–128.

20 Dawson, D. (2001) The private finance initiative: a public finance illusion? *Health Economics*, **10**, 479–486.

21 Pollock, A.M. (2004) *NHS Plc*, Verso, London.

22 Pollock, A., Shaoul, J., Rowland, D. and Player, S. (2001) *Public Services and the Private Sector: a Response to the IPPR*, Catalyst, London.

23 Sussex, J. (2001) *The Economics of the Private Finance Initiative in the NHS*, Office of Health Economics, London.

24 Department of Health (2000) *The NHS Plan. A Plan for Investment. A Plan for Reform*, The Stationery Office, London.

25 Department of Health (2000) *For the Benefit of Patients. A Concordat with the Private and Voluntary Health Care Provider Sector*, Department of Health, London.

26 Department of Health (2003) *Choice of hospitals. Guidance for PCTs, NHS Trusts and SHAs on Offering Patients Choice of Where, They Are Treated*, Department of Health, London.

27 Department of Health (2002) *Reforming NHS Financial Flows. Introducing Payment by Results*, Department of Health, London.

28 Department of Health (2003) *Payment by Results: Consultation Prepared for 2005*, Department of Health, London.

29 Street, A., Abdu, I. and Hussain, S. (2004) Would roman soldiers fight for the financial flows regime? The re-issue of Diocletian's edict in the English NHS. *Public Money and Management*, **24** (5), 301–308.

30 Martin, S. and Smith, P.C. (1999) Rationing by waiting lists: an empirical investigation, *Journal of Public Economics*, **71**, 141–164.

31 Council Directive 93/104/EC (1993) *Official Journal of European Community*, **L307**, 18–24.

32 MacDonald, R. (2004) How protective is the working time directive? *British Medical Journal*, **329**, 301–302.

33 Bloor, K., Maynard, A. and Street, A. (2000) The cornerstone of Labour's "new NHS": reforming primary care, in *Reforming Markets in Health Care Open* (ed. P.C. Smith), University Press, Buckingham, pp. 18–44.

34 Smith, P.C. (2005) Performance measurement in health care: history, challenges and prospects. *Public Money and Management*, **25**, 213–220.

35 Smee, C. (2005) *Speaking Truth to Power*, The Nuffield Trust, London.

36 Walley, T., Mrazek, M. and Mossialos, E. (2005) Regulating pharmaceutical markets: improving efficiency and controlling costs in the UK. *International Journal of Health Care Planning and Management*, **20**, 375–398.

37 Dusheiko, M., Gravelle, H., Jacobs, R. and Smith, P. (2003) *The Effects of Budgets on Doctors' Behaviour*. Technical Paper 26. Centre for Health Economics, University of York, York.

38 Tuohy, C. (1999) Dynamics of a changing health sphere: the United States, *Britain and Canada. Health Affairs*, **18**, 114–134.

39 Tuohy, C. (1999) *Accidental Logics. The Dynamics of Change in the Health Care Arena in the United States, Britain and Canada*, Oxford University Press, New York.

40 Enthoven, A. (1985) *Reflections on the management of the NHS*. Nuffield Provincial Hospitals Trust, London.

41 Enthoven, A. (1999) *In pursuit of an improving National Health Service*. Nuffield Trust, London.

42 Martin, S., Smith, P.C. and Leatherman, S. (2006) *Value for Money in the English NHS: Summary of the Evidence*. CHE Research Paper 18 University of York, York.

43 Office for National Statistics (2004) *Public Service Productivity: Health*, Office for National Statistics, London.

44 Holland, W. (2004) Health technology assessment and public health: a commentary. *International Journal of Technology Assessment in Health Care*, **20**, 77–80.

45 Stevens, A. and Milne, R. (2004) Health technology assessment in England

and Wales. *International Journal of Technology Assessment in Health Care*, **20**, 11–24.

46 Drummond, M. (2007) NICE; a nightmare worth having? *Health Economics, Policy Law*, **2**, 203–208.

47 Birch, S. and Gafni, A. (2007) Economists' dream or nightmare? Maximizing health gains from available resources using the NICE guidelines. *Health Economics, Policy Law*, **2**, 193–202.

48 Williams, I., Bryan, S. and McIver, S. (2007) How should cost-effectiveness analysis be used in health technology coverage decisions? Evidence from the National Institute for Health and Clinical Excellence approach. *Journal of Health Services Research & Policy*, **12**, 73–79.

49 Appleby, J., Devlin, N. and Parkin, D. (2007) NICE's cost effectiveness threshold. How high should it be?. *British Medical Journal*, **335**, 358–359.

50 Williams, A. (2004) *What Could Be Nicer Than NICE?* Office of Health Economics, London.

51 Heath, I. (2004) View of health technology assessment from the swampy lowlands of general practice. *International Journal of Technology Assessment in Health Care*, **20**, 81–86.

52 Birch, S. and Gafni, A., (1992) Cost effectiveness/utility analyses. Do current decision rules lead us to where we want to be? *Journal of Health Economics*, **11**, 279–296.

53 Bryan, S., Williams, I. and McIver, S. (2007) Seeing the NICE side of cost-effectiveness analysis. A qualitative investigation of the use of CEA in NICE technology appraisal. *Health Economics*, **16**, 179–193.

54 Coulter, A. (2004) Perspectives on health technology assessment: response from the patient's perspective. *International Journal of Technology Assessment in Health Care*, **20**, 92–96.

55 Linden, L., Vondeling, H., Packer, C. and Cook, A. (2007) Does the National Institute

for Health and Clinical Excellence only appraise new pharmaceuticals? *International Journal of Technology Assessment in Health Care*, **23**, 349–353.

56 Coast, J. (2004) Is economic evaluation in touch with society's health values? *British Medical Journal*, **329**, 1233–1236.

57 Dolan, P., Shaw, R., Tsuchiya, A. and Williams, A. (2005) QALY maximisation and people's preferences: a methodological review of the literature. *Health Economics*, **14**, 197–208.

58 Chinitz, D. (2004) Health technology assessment in four countries: response from political science. *International Journal of Technology Assessment in Health Care*, **20**, 55–60.

59 Sheldon, T.A., Cullum, N., Dawson, D., Lankshear, A., Lowson, K., Watt, I., West, P., Wright, D. and Wright, J. (2004) What's the evidence that NICE guidance has been implemented? Results from a national evaluation using time series analysis, audit of patients' notes, and interviews. *British Medical Journal*, **329**, 999–1003.

60 Freemantle, N. (2004) Commentary: is NICE delivering the goods? *British Medical Journal*, **329**, 1003–1004.

61 Stevens, S. (2004) Reform strategies for the English NHS. *Health Affairs*, **23**, 37–44.

62 Robinson, R. (2002) NHS foundation trusts. *British Medical Journal*, **325**, 506–507.

63 Smith, P. (2004) The scandal that persuaded ministers to "let go" of the NHS. *Health Service Journal*, **114**, 12–13.

64 Hauck, K. and Street, A. (2007) Do targets matter? A comparison of English and Welsh national health priorities. *Health Economics*, **16**, 275–290.

65 Yates, J. (2004) UK evidence on waiting, in *How Much Should We Spend on the NHS? Issues and Challenges Arising from the Wanless Review of Future Health Care Spending* (eds J. Appleby, N. Devlin and D. Dawson), King's Fund, Office of Health Economics and Centre for Health Economics University of York, London.

66 Wanless, D. (2002) *Securing Our Future Health: Taking a Long-Term View. Final Report*, HM Treasury, London.

67 Department of Health (various years) *Hospital Inpatient Waiting List Statistics, England, NHS Trust Based: the Green Book*, Department of Health, London.

68 Smith, P.C. and York, N. (2004) Quality incentives: the case of UK general practitioners. *Health Affairs*, **23**, 112–118.

69 Oliver, A. (2007) The Veterans Health Administration: an American success story? *The Milbank Quarterly*, **85**, 5–35.

70 Boyle, S. (2007) Payment by results in England. *Eurohealth*, **13**, 12–16.

71 Pedersen, K.M., Christiansen, T. and Bech, M. (2005) The Danish health care system: evolution – not revolution – in a decentralized system, *Health Economics*, **14**, S41–57.

72 Ryan, M., McIntosh, E., Dean, T. and Old, P. (2000) Trade-offs between location and waiting times in the provision of health care: the case of elective surgery on the Isle of Wight. *Journal of Public Health Medicine* **22**, 202–210.

73 Dawson, D., Gravelle, H., Jacobs, R., Martin, S. and Smith, P.C. (2007) The effects of expanding patient choice of provider on waiting times: evidence from a policy experiment. *Health Economics*, **16**, 113–128.

4
Finland

Unto Häkkinen

Population (millions)	5.2
GDP per capita (US$ PPP)	33 300
Health spending as % of GDP	7.5
Public health spending as % of total spending	77.8
Health spending per capita (US$ PPP)	2331
Acute care beds per 1000 population	2.9
Practicing physicians per 1000 population	2.4
Life expectancy at birth (years)	78.9
Infant mortality per 1000 live births	3

Strategies used

1. Decentralization
2. Information guidance
3. Strategic planning
4. Increasing size of municipalities
5. Regulation of wholesale drug prices
6. Generic substitution

4.1
Introduction

Before the 1990s cost containment was not an important issue in Finnish health care. The former planning and state subsidy system was an example of global budgeting which functioned satisfactorily during a period of steady growth in the economy [1]. Efficiency and cost containment became the aims of Finnish health policy during the early 1990s, which was much later than in many other countries [2]. Subsequently, the importance of the two topics in official statements has been rather similar, although the justifications for them have changed. During the early 1990s, Finland experienced a huge economic recession which required the cutting of public expenditure.

Cost Containment and Efficiency in National Health Systems: A Global Comparison
Edited by John Rapoport, Philip Jacobs, and Egon Jonsson
Copyright © 2009 WILEY-VCH Verlag GmbH & Co. KGaA, Weinheim
ISBN: 978-3-527-32110-0

At that time, efficiency in health care was assumed to be achieved by changing the structure of service. For example, in 1993 efficiency was stated as a health policy area where new polices must be created in response to new challenges. Efficiency was defined as ". . . improving health care services from the point of view of economy and effectiveness" [3]. Efficiency was assumed to increase as a result of a change in the structure of care brought about by transferring resources from institutional care to outpatient services, so that high-quality services were provided in a way that was reasonable from the point of view of economy.

During the 2000s, the justifications for an increased focus on better efficiency has been mainly stated to be the growing need for health and social services because of the aging of the population. The purpose of the ongoing municipal sector reform is to create a firm structural and financial basis within municipal services so that the organization and provision of services will be secured in the future. At the same time, the quality, effectiveness, availability, efficiency and technological change of services are taken into consideration [4].

In this chapter I will analyze the strategies for cost containment and efficiency in Finnish health care. First, I will describe the organization and funding of the system and trends in cost and financing. Details will then be provided of cost-containment strategies in the two main financing systems, and their effects evaluated. Finally, the experience of Finnish strategies will be summarized and evaluated from a broader perspective.

4.2
The Finnish Health Care System

4.2.1
Structure

In its institutional structure, financing and goals, the Finnish health care system is closest to those of other Nordic countries and the United Kingdom, in that it covers the whole population and its services are mainly produced by the public sector and financed through general taxation.[1] The Finnish health care system can be described as one the most decentralized in the world. Even the smallest of the over 400 municipalities (local government authorities) are responsible for arranging and taking financial responsibility for a whole range of 'municipal health services'. From an international perspective, another unique characteristic of the system is the existence of another public finance scheme – the National Health Insurance (NHI) scheme – which reimburses in part not only the same services as the first scheme, but also those services provided by the private sector. In addition to subsidizing the use of specific private health services, the NHI scheme also finances occupational and student health services and outpatient medicines (Figure 4.1).

1) A more detailed description of the Finnish
    health can be found in Refs [5–9].

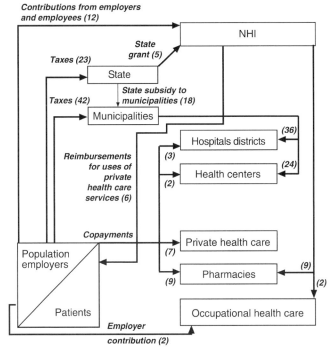

Figure 4.1 Main funding flows of the Finnish health system in 2003. Values in parentheses are percentages describing the shares of flow of total expenditure/financing.

Municipally provided services include primary and specialist health care. In addition, municipalities are responsible for other basic services, such as nursing homes and other social services for the elderly, child day care, social assistance and basic education. Municipal taxes, state subsidies and user charges finance the municipal health services. Primary health care is mainly provided at health centers, which are owned by municipalities or federations of municipalities. Preventive care for communicable and noncommunicable diseases, ambulatory, medical and dental care, an increasing number of outpatient specialized services, and various public health programs (e.g. maternity and school health care) are provided by the health centers. They also provide occupational health services and services for specific patient groups, for example clinics for diabetes and hypertension. Included within the health centers are inpatient departments in which the majority of patients are elderly and chronically ill, although n some municipalities the health centers also provide short-term acute curative inpatient services. In addition to the inpatient departments of health centers, long-term care is provided at homes for the elderly that are incorporated administratively under municipal social services.

Specialist care (both, psychiatric and acute nonpsychiatric) is provided by hospital districts, which correspond to the federations of municipalities. Each municipality must be a member of a hospital district. In addition to services provided through health centers or hospital districts, municipalities may purchase services from a private provider.

The second public financing scheme, the NHI, covers its members (i.e. all Finnish residents including people who are not working) in the following fields: sickness allowances; maternity allowances; special care allowances; student health services; rehabilitation services; and medical expenses (drugs prescribed by a doctor, private-sector examinations and treatments performed or prescribed by a doctor or dentist, and transportation services). In addition, it partly reimburses employers for the costs of occupational health services. The benefits of the NHI are financed mostly through compulsory contributions from insured persons (1.5% of income in 2005) and employers (1.6% of wages in 2005).[2] In principle, the NHI scheme is open-ended and the Government covers any deficit. Although the law defines the reimbursement system, the government and Ministry can to some extent effect the level of reimbursements by defining the basic tariffs and other details of the payment system.

4.2.2
Trends in Cost and Financing

Usually, health care spending is analyzed by the proportion of GDP devoted to health and health expenditure per capita.[3] The proportion of GDP spent on health services increased from 7.7 to 9.0% during the period 1990–1992; by 2000 it had decreased to 6.6%, and in 2005 it rose to 7.5%. The main reason for this exceptional trend was that the country experienced an unusually severe economic recession during the early 1990s when, between 1990 and 1993 the unemployment rate increased from 3 to 18% (and was 8% in 2001 and 6% in 2007). The per-capita GDP (at constant prices) decreased by over 10% during the space of these three years. Although some economic growth occurred during the two following years, in 1995 the per capita GDP was below the figures of the late 1980s. Thus, during the early 1990s the increase in the share of GDP can be explained by a greater decrease in GDP (Figure 4.2), although in the first year of the recession (1991) there was a 1% increase in the volume of health care (as measured by health expenditure at constant prices).[4]

2) The financing of NHI was reformed during early 2006. Health insurance was divided into work income and health care insurance. The latter covers drugs prescribed by a doctor, private-sector examinations and treatments performed or prescribed by a doctor or dentist, transportation services, student health services and rehabilitation services.

3) The figures used in this chapter are based on old statistics of health expenditure [10]. More recently, health expenditure in Finland has been calculated according to SHA definitions [11]. The new calculations suggested an increase in health expenditure in Finland by 5% in 2005. The share of GDP devoted to health care will increase from 7.5 to 8.3%.

However, the figures are not yet available from earlier years and thus are not used here. In addition, to date only a few countries have implemented SHA in their figures.

4) It should be noted that price indices for GDP and health expenditure are developed using different starting points, which makes the comparison of prices and volumes somewhat difficult. The GDP price index describes the general trend on output prices, whereas health care price index (at least in Finland) describes the trend in input prices which reflects very much the development of wages in the health sector. Thus, development in productivity is included in the GDP price index but not in the health care price index.

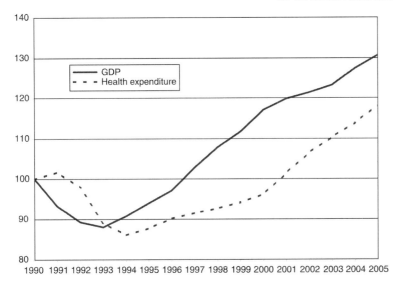

Figure 4.2 Per-capita health expenditure and GDP at constant prices in Finland 1990–2005 (index 1990 = 100).

During the period 1991–1994, the volume of health care per capita decreased by 15%. Following this the GDP rose again, thereby further decreasing the proportion of GDP spent on health services. At constant prices, per-capita health care expenditure reached the level of the early 1990s in 2001.

From an international perspective the level of health expenditure [measured both in terms of GDP share and PPP (purchasing power parity) -adjusted expenditure per capita] is much lower than in other OECD countries. This can be partly explained by fact that the relative wages of health care personnel are considerably lower in Finland than in other countries [5], which has not been taken into account in international comparisons of health expenditure.

During the 1990s the changes in health care financing (Table 4.1) were mainly due to two factors, which were also attributable to the poor economic situation. First, the amount of state subsidies for municipal services fell, meaning that the municipalities assumed – in addition to an increased freedom of choice regarding the provision of services (see also below) – greater economic responsibilities for providing these services. Thus, the share of health care financed by municipal taxes increased. One main reason for the decrease in state subsidies was the steep increase in unemployment expenditure and the decrease in tax revenues in the state budget. This resulted in a large increase in net Government borrowing.

Second, the share paid by households in financing health care increased from 13 to 20% between 1991 and 1993. The official reason [12, 13] for the increase was that it encouraged a more effective allocation of service use and enabled an increase in public funding. The increase in cost sharing stemmed partly from the abolition of a

Table 4.1 Financing of health care between 1990 and 2005 in Finland.

	Share (%)			
	1990	**1995**	**2000**	**2005**
State	37.2	28.4	18.4	20.7
– direct taxes	14.1	5.5	8.6	8.6
– indirect taxes	22.1	12.5	9.8	12.1
– net borrowing	1.0	10.4		
Municipalities, direct taxes	35.8	33.8	41.3	40.3
National Health Insurance	10.8	13.4	15.4	16.6
– contributions from employees	4.9	8.7	6.9	4.7
– contributions from employers	4.8	4.7	5.9	4.8
– state	1.1		2.6	7.1
Public finance, total	83.8	75.6	75.1	77.6
User charges	12.6	20.5	20.4	17.9
Other sources	3.6	3.9	4.5	4.5
Total	100	100	100	100

tax deduction for medical expenses from income taxes in 1992 and partly from the increase of user charges for municipal health services in 1993.

In this Chapter I analyze the specific strategies adopted for cost containment in Finland in the two main funding systems. As indicated above, much of macro-level trends in Finnish health can be explained not only by external factors (such as economic factors) but also by specific organizational and institutional features [6, 7]. This means that it will be very difficult to separate the effects of specific cost-containment strategies from effects of external factors. In addition, because of the high degree of decentralization in Finnish health care, most important cost-containment strategies have been made at the municipal level – which means that there are over 400 different strategies. Here, I will describe the strategies adopted by the central government as well as municipalities and hospital districts in terms of the actions that they have adopted.

4.3
Strategies for Municipal Health Services

The cost-containment strategies for municipal health service have been mostly focused on the relationship between the state and municipalities (Table 4.2). Since municipalities are responsible also for public services other than health care, most important strategies have not yet been directed towards health policy but more generally towards the relationship between state and municipalities. Within this general framework the government (Ministry of Health and Social affairs) has adopted its own strategies (information guidance, strategic planning), the aims of which have also been to improve efficiency and contain cost.

Table 4.2 The main reforms and changes in the municipal health
services in Finland during the 1990s and 2000s.

Year	Reform/Change	Aim of the reform/Change
1993	State subsidy reform. Reduction of central government control and increase in local freedom in the provision of services	Increase efficiency through decentralization
1993	Increase of user charges in municipal health services	Contain public expenditure
1993 onwards	Information guidance and Government's strategic planning	Increase efficiency and contain cost
Late 1990s and 2000s	Local projects and experiments	Contain costs
2002 onwards	Securing the Future of health care (e.g. introducing waiting time guarantee and increase of finding to FinOHTA)	Ensure accessibility, time availability and high-quality health care services
2005 onwards	A project on restructure municipalities and services. Decreases the number of municipalities responsible for providing health services	Contain cost

4.3.1
Decentralization of Responsibilities

The most important reform in Finnish health care during the past decade occurred in early 1993 as part of a reform of the entire state subsidy system. An essential element of the reform was the revision of the grounds for determining state subsidies to municipalities for health services. Under the old system, state subsidies to municipalities or federations of municipalities (producers) were earmarked and related to real costs. Under the reformed system, however, state subsidies for running costs in health services provided by municipalities are nonearmarked lump-sum grants, which are calculated prospectively by using a specific needs-based capitation formula. For health services, these subsidies are calculated according to certain criteria; during the period 1993–1996 these included population, age structure, mortality [standardized mortality ratio (SMR) for all ages], population density, land area and the financial capacity of the municipality. The archipelago and other remote area municipalities received a somewhat higher subsidy. New criteria were adopted from the beginning of 1997 onwards which included population, age structure and an age-standardized index of invalidity pensions for persons under the age of 55 years.

The aim of the Finnish reform was to reduce central government control and to increase local freedom in the provision of service. Thus, the aims of the reform can be seen in the light of the framework of fiscal federalism [14]. Its main argument is that public goods consumed locally should be produced locally. Decentralization is believed to lead to increased welfare by allowing local authorities to act in accordance with local preferences and local cost structures. In Finland, local preferences are assumed to be included in the provision of services through local elections. In addition, since municipalities can also decide the level of local taxes they were assumed to have incentives to behave in an efficient way that will also contain the cost of services. The incentive issue was greatly intensified since the average proportion of the state subsidy in the total health budgets of the municipalities dropped from 41 to 25% between the years 1991 and 1998.

The reform of 1993 changed the financing of hospital care. Before the reform, the state subsides were given directly to producers (hospitals), but after the reform they were given to municipalities. Thus, the hospitals obtain their revenues by invoicing the municipalities. As purchasers, the municipalities negotiate the provision of services with their hospital district on an annual basis. Local politicians are involved on both sides of the purchaser–provider relationship: they are decision-makers on the elected municipal council and also in the hospital district and hospital administration.

However, within the field of specialized hospital care, the asymmetry of information between the providers (hospitals and hospital districts) and the buyers/financiers (municipalities) is substantial, particularly in the case of the small municipalities. The latter are also economically weak in comparison to the large hospital district authorities; most of the over 400 municipalities are also too small to pool specialized health services, because of the associated financial risk [15, 16].

Only rarely is there a long-term incentive for a municipality to buy services from other hospital districts or a private hospital, because this would undermine the financial situation of its own local hospital.[5] In the absence of nationally set guidelines, hospital districts determine the prices for their services, and the method by which services are defined and prices calculated varies from district to district [16]. The pricing of hospital services has been in a continuous state of flux, and consequently the opportunities for municipalities to compare prices are very limited. Competition is also restricted by the fact that a hospital district is a local monopoly in its area since, according to the law, a municipality must be a member of a hospital district.

In hospital pricing the current trend is away from the old bed-per-day price towards case-based prices. The main reason for this has been to make the financing of hospital care between municipalities more equitable – that is, to better reflect the real cost of care given to patients. In 1993, for example, 7% of Finnish hospitals invoiced hip replacement using case-based prices, whilst 64% had done so by 1998. The trend towards case-based prices has been similar for many other procedures, although a little slower. In 1997, a large southern hospital district introduced the first hospital invoicing systems relying on diagnosis-related group (DRG) case-based pricing. In

---

5) The only exception so far is a highly specialized hospital (Coxa) that was founded in Tampere in 2002 to perform endoprosthetic operations. This functions as a limited company, and was founded by Pirkanmaa hospital district (and three other hospital districts), four cities, one Finnish foundation and a German private hospital company. In 2005 the German com- pany sold all its shares to the Finnish National Fund for Research and Development. All elective endoprosthetic operations of Pirkan- maa hospital district are carried out in Coxa hospital. It also provides these services for patients from other hospital districts, as well as private patients.

2005, nine out of 21 hospital districts and 15 out of 42 hospitals used DRGs. These nine hospital districts produced about 50% of all specialist services in the country, with 43–75% of the total payments being based on DRGs. Thus, about 30% of the expenditure on somatic specialist care in Finland is based on DRG payments. In addition, the Helsinki-Uusimaa hospital district has begun to develop DRG-based pricing for outpatient and psychiatric care [17].

4.3.2
Governments Strategic Planning

The state subsidy system was implemented during the huge economic recession in Finland. This reform offered an option for advancing deregulation, and distanced the national government and health minister from the unpopular implementation of cost-containment measures and budget cuts necessitated by the economic recession. In terms of political accountability, decentralization allowed the central government politicians to deflect some of the blame for inadequacies in health care provision to the local governments. At the same time, the local politicians could blame the central government cuts in state grants for health care for the same inadequacies.

Following the state subsidy reforms, the strong state regulation (such as a firm control over the personnel employed and the mix of personnel) changed to a '*softer regulation*' or '*information guidance*' within the system. This relied on the assumption that the provision of information to municipalities, producers and professionals (doctors) would drive any constructive behavioral or system change. The information guidance included aspects such as: improving the statistical systems to allow more transparency concerning costs, outputs, accessibility and effectiveness of the different municipalities and service providers and comparisons between them ('benchmarking'); producing information to support 'evidence-based' choices of effective technologies and practices in health care (e.g. the establishment of FinOHTA – the Finnish Office for Health Care Technology Assessment – to guide the choice of technologies in health care); creating a continuing education program aimed at more rational drug therapy; and developing national nonbinding recommendations on personnel and other resource requirements and practices for service provision.

Under current legislation the power of the ministry is very weak, and it does not have any effective means to affect decisions made at the local level. The national regulation is limited to legislation only, and oversight of municipal health services is mainly in response to complaints or other highly visible problems in the operation of the services. If the state level administration (the ministry or other, state authorities) detect overt violation or neglect of existing health service legislation, they can intervene. Usually, this means raising problems to start a discussion, or issuing reminders or formal warnings. Since the mid 1990s the government has – in addition to information guidance and subsequent monitoring – implemented so-called 'strategic planning'. This includes strategic policy documents, a number of working parties, committees and development projects, as well as the hiring of experts to develop recommendations for actions to municipalities within the field of health care. Some of the development projects have also included regional committees in order to engage local decision makers to action. Cost containment has usually been mentioned as one of the main starting points for these actions and strategies.

The results of many working parties, committees and development projects have been a large number of recommendations. Unfortunately, however, the effect of these actions has been minimal. One reason for this has been the diversity of recommendations; the national developmental work and projects do not become concrete in practice at local level. It has also been argued that the municipalities (i.e. local politicians) do not have enough incentives and expertise to develop the structure and content of health services [18].

During recent years, the Government's involvement in the provision of health care has increased. These actions have included extra earmarked funding to municipalities and hospital district for certain tasks (to increase psychiatric services for children, to reduce the number of patients on waiting time and to shorten waiting times). In 2001 the Government initiated 'the National Project to Ensure the Future of Health Care', which was originally proposed by the Prime Minister and the Minister of Social and Health Services at the time. The project aimed to solve a variety of deficiencies identified in the Finnish health care system. The main outcome of the project was a working group memorandum and "... the Decision in Principle by the Council of State on Securing the Future of Health Care" issued by the Government in April 2002 [19]. This focused on strengthening primary health care and preventive work, ensuring access to treatment, ensuring the availability and expertise of personnel, reforming of functions and structures and augmenting the finances of health care. As a result of the decisions that were made based on the project, the annual government funding of FinOHTA has increased from €1.1 million to €2.2 million between the years 2004 and 2007. Nowadays, FinOHTA coordinates health technology assessment (HTA) research, disseminates information and provides methodological and financial support to research projects (aimed at evaluating the clinical efficacy or cost-effectiveness of a given health technology). An external review of FinOHTA recommended that also in the future, it should continue developing and focusing its mission and position as the national coordinator, facilitator and expert in technology assessments [20]. Between 1995 and 1998, FinOHTA has grown from a small unit of three to five employees to a considerable larger unit; in 2007, the FinOHTA staff was about 30 persons, covering a variety of professional expertise.

In March 2005 the Government began to implement two reforms. The first reform was to introduce clinical guidelines for a wide range of treatments, aimed partly at bringing about some convergence across Finland in rates of elective surgical procedures and in the setting of thresholds for admission to waiting lists for procedures. The second reform was to introduce a set of maximum waiting-time targets for nonurgent examinations and treatments at health centers and hospitals. Recommended hospital treatments, including elective surgery, should be offered within three to six months; however, if the patient's own health center or hospital is unable to provide the necessary assessment or treatment within the set time frame, then it must arrange the option of treatment in another municipality or in private health care, without extra cost to the patient.

4.3.3
Cost Containment at the Local Level

The state system reform in 1993 made it possible for municipalities to adopt a more active role as a purchaser, instead of acting in the mainly producer's role as previously. In particular in the field of specialist hospital care, the reform meant that the system had changed somewhat from a public-integrated model to a public-contract model. The reform gave the municipalities and hospital districts enormous freedom to organize, regulate and administer service provision. It also extended the right of the municipalities to purchase services freely from public, not-for-profit and for-profit providers and informal caregivers, and to contract out existing public services. The deregulatory part of the reform included dismantling a number of legal and administrative norms applying to the administration, personnel and user charges of municipalities and health care providers.

The municipalities got their increased responsibility for health care in the time of economic recession, which meant that cost containment and even the cutting of cost was the main starting point for their actions. It is widely considered that this experience of saving and focusing on cost containment much affected the behavior of the municipalities during the late 1990s and even the early 2000s, when the economic situation of the municipalities and the whole country was better.

In the 1990s, during and after the time of economic recession, the municipalities did not use their increased power to reorganize the services or purchase services from private sector. During the period 1993–1997, the share of municipal expenditure devoted to purchasing private services decreased, even though the reform program now permitted the municipalities to contract-out services [21]. Only rarely was there a long-term incentive for a municipality to buy services from other hospital districts or a private hospital, because this would undermine the financial situation of its own local hospital or health center. The health sector is also an important employer, and its employees generate income tax revenues for the 'host municipalities' of the providers. Therefore, some municipalities have been willing to pay more for services provided by a hospital or health center located in their municipality.

During the 2000s, local authorities have begun to reorganize models of production. In general terms, the aim of these local projects has been to contain cost by co-operation both between different sectors within municipal health services (primary care, specialist care and care of the elderly) and with the private sector. In some current 'privatizations' a primary reason has also been the difficulties to recruit a work force (mainly physicians), and local projects and experiments have emerged in quite different directions. These projects have included the development in one hospital district of a clearer purchaser–provider model in which smaller municipalities have formed cooperative purchaser organizations for arranging specialized services; a municipality buying its health services from the nonprofit third sector; and an instance of the merging of health centers and a district hospital into a single organization providing all health services for inhabitants of the municipalities in the area.[6] Some hospital districts have transformed laboratory services as publicly owned companies, which can then provide services to the hospital district, municipalities and also to the private sector. In addition, some nonmedical services are

6) The most important are the following [9]. In 2001, five municipalities (the most populated being Forssa) in southern Finland, with a population of 36 000, founded a common organization to provide primary and specialized health services. In practice, health centers and one regional hospital was merged to this organization. The regional hospital was withdrawn to the new organization from the hospital district of the area (Kanta-Häme). Municipalities were cooperating in providing primary health care already before the reform. The main goal of the reform was to keep the municipalities' cooperation functioning, to keep the decision-making regarding specialized services as local as possible, and to enhance cooperation between primary health care and specialized health care. The reform has thought to be generally very successful.

The new administrative experiment in the Kainuu region (North-East Finland) started in 2005. This covers nine municipalities having a total of 85 000 inhabitants. The experiment created a new regional, self-regulating mid-level administrative body with its own regional council elected for a four-year term, at the same time as the general municipal councilors' election. The county has no right to levy taxes, but obtains its funding from government subsidies and from municipalities. In this experiment, the county level municipal federation took responsibility for several welfare services which were previously run by the municipalities: upper secondary schools and vocational education; primary health services; specialized health care; and a large part of social services. In this experiment, the provision of primary health care and specialized health care (municipal health centers and Kainuu central hospital) were merged to the same organization. Among other things, this has provided a possibility of unifying the electronic patient record systems.

The health center and district hospital in the municipality of Mänttä was merged into a single organization that provides all health services for the inhabitants of Mänttä and the nearby municipality of Vilppula in 2002. Mänttä and Vilppula have together a population of about 12 000. This means that both the primary and specialized care is arranged by

one organization. The new organization was created as one subunit of the Pirkanmaa hospital district. The aim of the new model is to secure health services for the local population at reasonable cost.

From the start of 2005 the joint health center of Pietarsaari and two neighboring municipalities joined the regional hospital situated in Pietarsaari to form 'Malmin terveydenhuoltoalue' (health care district). These municipalities had had a common health center since 1973. Before the reform, the regional hospital was part of the hospital district. The municipalities together have a population of about 34 000.

Two most recent reforms of this type are conducted in the Itä-Savo and Päijät-Häme regions. In both regions the municipalities formed new organizations to provide primary and secondary care and social services (started on 1st January 2007). The new organizations replaced hospital districts which provided only secondary medical services. Like hospital districts, the new organizations are municipal federations that are governed by member municipalities.

Itä-Savo district is located in eastern Finland, and has nine municipalities as members (a population base of 60 000). One of the municipalities is a small city, while the others are small rural municipalities. All member municipalities purchase secondary care services from the new organization; seven of the municipalities purchase primary health care services (80% of population); and three of the municipalities also purchase some social services such as elderly care and services for alcohol and drug abusers (62% of the population). The district has eight health stations and one hospital.

Päijät-Häme district is located in southern Finland and has 15 municipalities as members (total population of 210 000 inhabitants). One of the municipalities (the city of Lahti) is the seventh largest city in Finland. The new organization is responsible for providing secondary care services for all the member municipalities, and primary health care and social welfare services for eight member municipalities having a total population of 51 000.

outsourced in some hospital districts (e.g. catering and laundry services), and cooperation with the private sector has also been increased.[7]

4.3.4
Increasing the Size of Municipalities

The state subsidy reform gave the responsibility for health care to municipalities. The median population of the municipalities is 4700; the smallest have fewer than 1000 inhabitants, while large municipalities such as the cities of Helsinki and Tampere have about 550 000 and 200 000 inhabitants. Thus, even the smallest municipalities are responsible for arranging and taking responsibility for a whole range of health services. During the 1990s many recommendations were made in order to increase the population base for units responsible for health services. However, the effect of these has been minimal, and even an opposite trend can be observed. For example, since 1993 there has been a clear tendency for dissolving health center federations, and even small municipalities have decided to produce services themselves.

As a response to the fact that municipalities do not want to cooperate voluntarily, in February 2005 the Government established a project to restructure municipalities and services. The background was the concern about the increasing financial difficulties faced by municipalities, and the growing need for health and social services due to aging of the population. The latter effect would greatly reduce the availability of the workforce, as a remarkable number of personnel within social and health services would retire during the next decade or so. But, at the same time, it would increase the need for the workforce.

The purpose of the planned public sector reform is to create a firm structural and financial basis within municipal services so that the organization and provision of services will be secured in the future. At the same time, quality, efficacy, availability, efficiency and technological change of services are taken into consideration. The project concerns all services organized by municipalities – not only health care – and the expected outcome is a restructuring of the municipalities and services.

The project has made three different proposals for organizing basic services: a model of basic municipalities; a regional model; and a district model. The first model would merge the smallest municipalities into larger ones with a minimum population size of 20 000 inhabitants. The regional model would introduce 20 municipalities with a relatively large population size and responsibilities similar to those of the

---

7) Examples can be found in Karjaa and Lahti [9]. The municipality of Karjaa agreed with Samfundet Folkhälsan (nonprofit, 'third sector' organization) in 1998 that it would purchase all primary health care and geriatric services from Folkhälsan. To provide the services Folkhälsän founded a company which it owns in its entirety. Karjaa rented the facilities used for providing the services to this company. In 2004, the City of Lahti made a contract with a Finnish private company MedOne to provide all services of one of its health stations. The business activity of MedOne concentrates on the outsourcing of health care services, mostly to lease health care professionals (mainly physicians) to public health care. The personnel of the health station were transferred to this private company. More recently, health stations from, for example the city of Kotka and city of Kouvola, have been outsourced to MedOne.

current municipalities. The third model would integrate primary and secondary health care as well as certain social welfare services into one and the same organization, with a population size of 100 000–200 000 inhabitants, while leaving the responsibility for the remaining basic services to current municipalities.

In January 2007, Parliament accepted a skeleton law about how to continue the process. According to that act, the Government will provide financial support for the mergers of municipalities. The act also states that primary health care and social services closely related to health services should be organized by organizations covering at least 20 000 inhabitants. This would not necessarily require the merging of municipalities with fewer than 20 000 inhabitants, but would form for example municipal joint federations. Currently, only about one in four health centers has a population base of 20 000 or more, so the municipalities had to plan how these goals would be achieved during the year 2007. Although the issues of organizing hospital services were on the agenda, during the early stages of the project the skeleton legislation does not include much new information. Rather, it confirms the early situation: the responsibilities of services that require a large population base are given to federations of municipalities (e.g. hospital districts) to the extent determined by municipalities. At the time writing of this Chapter, 44 municipalities has decided to merge, so that total number of municipalities will have decreased by 31 by the start of 2009. An additional 90 municipalities are considering such a merger.

The *centralization of hospital care* to larger units has also been under discussion, and has been recommended by the Ministry of Health; however, this has not been supported by either legislation or regulation. Some centralization has occurred on a voluntary basis; for example, in 2007 the merger of three hospitals (Helsinki University Hospital, and Jorvi and Peijas Hospitals) in the capital area will create one very large unit that will be responsible for about 25% of all acute somatic care in the country (in monetary terms). As the new unit will be organized under medical specialties, the same specialties of the former three hospitals were merged.

4.3.5
Evaluation of the Strategies

The Finnish decentralized model is likely to have both positive and negative effects on efficiency. The OECD [5] concluded that the potential 'pros' included the gains from allowing communities to exercise local preferences and use local knowledge, while potential 'cons' included problems of variations in taxable capacity, losses from reduction in purchasing power, losses of economies of scale, lack of expertise and conflicts of interests.

The effects of *decentralization* are very difficult to analyze, as it was implemented during the time of economic recession. However, it seems that the Finnish model has contained expenditure quite well. The GDP share of the municipalities' expenditure on health care and nursing homes decreased between years 1993 and 2000, from 5.2% to 4.3%. Subsequently, it has increased such that the share was 4.9% in 2005. This development which occurred during the 2000s can be partly explained by increases in the wages of doctors following their strike in 2001 and an increase in state subsidies in recent years. At constant prices, the municipalities' per-capita expenditure on health care and nursing homes has increased between 1993 and 2005 from €1170 to €1472 – that is, an annual increase of about 1.9% (Figure 4.3). In 2000s, the annual increase was 3.2%, although only a small part of this was due to changes in demographic factors. Age, gender and needs standardized expenditure[8] were seen to increase at 1.5% per annum during the whole period, and at 2.8% during the years 2000 to 2005.

Regional inequity is usually considered to be a major disadvantage of decentralization. In Finland, regional differences in adjusted expenditure per capita on health care and nursing homes have not increased since the reform. In fact, variation coefficients have even indicated some decrease in municipal variation between the years 1993 and 1998. In 2005, the coefficient was lower than in 1993 (Figure 4.3), although the variation was still considerable. In 2005, the variation between the extreme municipalities was €1000–2000 per capita. When municipalities expenditures are aggregated at the hospital district level, the level of per capita expenditure in the cheapest area in 2005 was similar to the mean of the country in 1993, and still much lower than in the most expensive area in 1993.

Some evidence exists that changes in productivity[9] in Finland are more closely associated with direct economic constraints (affecting municipalities) than with decentralization or changes in incentive (financial) structures. Among health centers, there was a substantial decline in productivity from 1988 to 1990, but a clear increase during the period 1991–1995 [23]. Thus, the upturn in productivity occurred at the same time as the municipalities suffered financial problems due to the recession, as well as the change in financial incentives. The same trend was also found in a study on hospital productivity during the period 1988–1994, there being a significant increase in productivity during the years 1991–1994. However, in this case, much of the observed increase in productivity was due to advances such as the introduction of day surgery and other new technologies that reduced the average length of stay. In addition, the greatest increase in productivity occurred during the early 1990s, during the time when the hospital funding, pricing and incentive systems remained largely unchanged [24]. Studies conducted during the late 1990s and the early part of the current decade (i.e. at a time of increasing funding) have indicated a decreasing trend in productivity, both among health centers and in acute somatic hospital care [25, 26].

Those studies that have related health expenditure or productivity to the size of a municipality have not shown any clear evidence of economies of scale [27]; (L. Nguyen *et al.*, unpublished results). However, a recent study monitoring the

---

8) The age, gender and needs standardization is based on study on developing the formula for state subsidies for health care in Finland [22]. The aim of standardization is to take into account the differences between municipalities (and in this case also between years) in age, gender and mortality (measured by factors such as mortality and index of disability pensions) structure of the population.

9) Productivity here means the ratio of outputs to inputs, where outputs are measured in terms of output indicators or intermediate outputs such as discharges, visits, bed-days, procedures, etc.

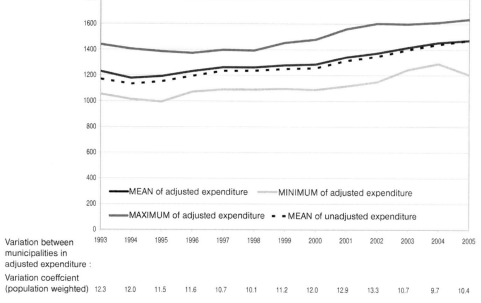

Variation between municipalities in adjusted expenditure :	1993	1994	1995	1996	1997	1998	1999	2000	2001	2002	2003	2004	2005
Variation coeffcient (population weighted)	12.3	12.0	11.5	11.6	10.7	10.1	11.2	12.0	12.9	13.3	10.7	9.7	10.4

Figure 4.3 Municipal age, gender and needs adjusted (and not adjusted) expenditure on health care and nursing homes per capita over the period 1993–2005 at constant (2005) prices. Data shown are mean, minimum and maximum by hospital district.

effect of the secession of municipal health center federations gives support to the larger units [28]. As mentioned earlier, primary health care is mainly provided at health centers, which are owned by municipalities or federations of municipalities. Usually, the smaller municipalities provided service by federations, but since 1993 there has been a clear tendency for the dissolution of former health center federations. In fact, between 1993 and 2003 a total of 37 such dissolutions occurred. According to the results of this study, the per-capita primary health care expenditure growth was approximately 8% higher in seceded health centers compared to (matched) nonseceded health centers. In addition, the secessions had no positive effects on productivity in the long term. The rapid expenditure in the growth of seceded health centers can be explained by both increasing service volume and decreasing productivity.

In *hospital pricing* it might be assumed that the earlier-described change towards case-based prices would have effects on the volume of services as well as on the length of stay, as they alter producer incentives (towards a more activity-based funding); these effects have already been observed in other countries (e.g. the USA and Sweden). However, a Finnish study [16] on the effects of case-based pricing in three common surgical procedures using panel data from 1991–1998 did not identify any clear evidence for this hypothesis. The use of case-based pricing increased the number of lumbar discectomies by 8% and decreased the length of stay for hip and

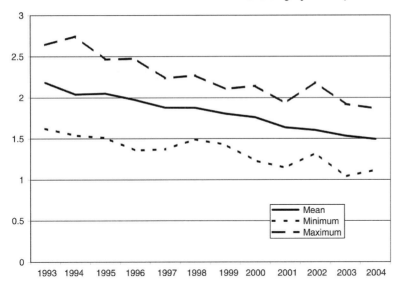

Figure 4.4 Age- and gender-standardized avoidable deaths in
Finland and their variation between hospital districts over the
period 1993–2004. Deaths are per 1000 population aged under
75 years.

knee replacements by about 0.5 days. On the other hand, case-based pricing did not
increase the volume of hip and knee replacements, neither did it decrease the length
of stay for lumbar discectomies. Although this study considered only three procedures, the results indicated that the reformed hospital financing methods do not
necessarily promote a more efficient resource allocation, as case-based pricing does
not materially increase the number of hip and knee replacements, which are
generally considered cost-effective. In addition, case-based pricing has directed
scarce resources towards increasing the number of lumbar discectomies, where
cost–effectiveness has not been clearly verified.

It should be noted that none of the above-referred studies was able to take into
account the changes in quality of service, measured for example by means of health
outcome indicators. One option is to look at trends in 'avoidable deaths', which can be
considered as a health output or quality measure. This is based on a list of causes of
death that should not occur in the presence of effective and timely health care [29].[10]
Since 1993 the avoidable deaths have decreased in Finland, but again the regional
differences have not changed (Figure 4.4) [30]. However, the age- and gender-
standardized mortality level in the worst hospital district in 2004 was clearly lower
than average level in the whole country in 1993.

---

10) The weakness of the indicator is that it is
always based on some extended arbitrary
choice about which deaths are avoidable. In
addition, the concept has limitations relating
to the comparability of data, attribution of
causes, and coverage of the range of health
outcomes.

The first results of an ongoing PERFECT[11] project also indicated clear regional and hospital differences in costs and outcomes in the care of very preterm infants [31], acute myocardial infarction [32], stroke [33] and hip and knee replacements [34]. Although, the reason for the differences have not yet been carefully investigated, a study of the care of very preterm infants has shown that the centralization of care to the neonatal intensive care units in the five university hospitals would reduce the one-year survival of infants [35].

4.4
Cost Containment in Services Covered by NHI

In spite of rather radical decentralization in the municipal health service the changes in the other public financing scheme (NHI) has been rather minor. The general aim of reforms here has been to contain the public expenditure. For example, the reform in the reimbursement system for occupational health by the NHI in 1995 aimed to promote preventive activities and contain costs. Maximum sums (per employee) for both compulsory and voluntary curative refunds were defined, and refunds for specialist services were limited. Moreover, refunds for family member utilizations were abolished. In other sectors reimbursed by the NHI (doctor's services, examinations and treatments) the cost has been contained quite effectively by considerable higher copayments than similar services given in municipal health services.

The NHI is the only public financier for medicines given outside inpatient care. During the past decade, expenditure on pharmaceuticals has increased more rapidly than other health expenditure or GDP. Reimbursement for prescribed drugs represents over 60% of NHI spending. The increase of NHI reimbursements for medicines has been the main reason why, since 1998, central government has been obliged gradually to increase its direct funding for NHI (see Table 4.1). Next, attention will be focused on various strategies that have been used to contain medicine expenditure.

4.4.1
The Finnish Pharmaceutical System

The Social Insurance Institution is responsible for the public financing of prescribed medicines through the NHI scheme, where it (in 2005) refunds 50% of all medical expenditures in excess of a fixed minimum sum per purchase (Basic Refund Category) or, in special cases, nearly all medical expenditures (registered individuals suffering from certain specified conditions, as decided by government, qualify for a 75 or 100% reimbursement in excess of a fixed minimum). The Lower Special

11) The PERFECT (PERFormance, Effectiveness and Cost of Treatment episodes) project has developed protocols for seven diseases (acute myocardial infarction, hip fracture, breast cancer, hip and knee replacements, very low-birth-weight infants, schizophrenia, stroke). Register-based indicators (both at regional and hospital level) on content of care, costs and outcomes are now available for the years 1998 to 2005.

Refund Category (75%) consists of 10 chronic illnesses (e.g. long-term hypertension, asthma, cardiac insufficiency), while the Higher Special Refund Category (100%) consists of 34 chronic illness (e.g. diabetes, cancer) where drug treatment is necessary and effective to maintain the patient's health status, and where the drug restores or replaces normal bodily functions. In addition, all outstanding prescribed medical expenditures in excess of a certain sum each year will be paid (Additional Refund). Thus, there is an annual expenditure ceiling for the prescribed medicines: if a patient paid more than €617 (in 2006) then NHI covered the entire drug cost, and the patient paid €1.50 per medicine per purchase.

Figure 4.5 broadly outlines the structure of decision-making within the pharmaceutical sector in Finland [36]. The Ministry of Social Affairs and Health sets the long-term strategic goals for the pharmaceutical policy and prepares the laws concerning medicines. Two official bodies under the ministry – the National Agency for Medicines (NAM) and the Pharmaceuticals Pricing Board – are the major decision-making bodies affecting the pharmaceutical markets. The NAM tightly controls the number and location of pharmacies (for private profit-making) licensed to sell medicines. Although the Social Insurance Institution has a say in the reimbursement process, it is not a decision-maker. The Finnish drug reimbursement system is based on a list of products that can be prescribed.

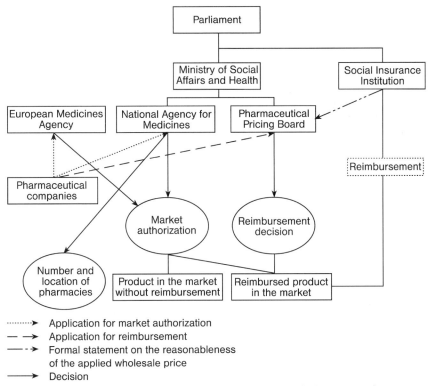

········▶ Application for market authorization
— —▶ Application for reimbursement
——-▶ Formal statement on the reasonableness
of the applied wholesale price
———▶ Decision

Figure 4.5 A simplified presentation of decision making in the Finnish pharmaceutical sector.

Before 1994, pharmaceutical products in Finland could only be sold with the permission of the NAM, and retail prices of medicines distributed by the pharmacies were regulated. After the implementation of the European Economic Area agreement at the start of 1994, a pharmaceutical with valid market authorization may be marketed without its wholesale price having been approved by the authorities. Companies can set their wholesale prices freely, but if they want to have the product reimbursed by the NHI scheme they are obliged to apply for reimbursement from the Pharmaceuticals Pricing Board. One criterion for a medicine to be approved for reimbursement by the Social Insurance Institution is that its wholesale price is reasonable. After the reasonable wholesale price has been approved, the product will automatically qualify for reimbursement under the Basic Refund Category. The Pharmaceuticals Pricing Board determines the 'reasonable level' of the wholesale price. 'Reasonable wholesale price' refers to the maximum price at which the product may be sold to pharmacies and hospitals. An assessment of the reasonable level of the price is based on four factors: (i) an economic evaluation of the product compared to major competitors or prevailing treatment practices; (ii) the wholesale price compared to major competitors and the price of the product in other EU countries; (iii) budget impact on the NHI; and (iv) on clinical judgment. It also partly determines the retail price on which the reimbursement is based. The retail price includes the pharmacies' profit margin, VAT and the pharmacy fee, all of which are defined by the Government.

On the whole, the decision-making for the wholesale prices of pharmaceuticals, as well as the number of pharmacies, is highly centralized. The NHI scheme reimburses pharmaceutical products on each inhabitant independently – that is, the reimbursement is not interlinked or co-coordinated with other health services, of which most are provided by municipalities and financed mainly by municipalities through local taxes and state subsidies.

4.4.2
Cost-Containment Measures

4.4.2.1 Cost Sharing and Price Control

The Finnish drug reimbursement system has been revised in several ways since the 1990s in order to curtail the growth in NHI expenditure (Table 4.3). During the early 1990s the main instrument was to increase patient cost sharing, whilst at a later date price control became the key measure.

The Working Group on Medicines Cost appointed by the Ministry of Social Affairs and Health, which was active during 1996 and 1997, proposed in its concluding report several measures designed to curb expenditure on medicines. Most of these measures were implemented in 1998–1999, and many of them (such as the abolition of compulsory stock billing, reviewing the whole prices of medicines in force and increasing the power of the Pricing Board) concerned the determination of a reasonable wholesale price.

After 1998, the Pricing Board could decide to reduce the valid wholesale price, if indications of the medicine were more extensive as compared to the situation when

Table 4.3 The main reforms and changes in the pharmaceutical markets in Finland during the 1990s and 2000s.

Year	Reform/Change	Aim of the reform/Change
1990	Fixed deductible for drugs in BRC increased from FIM 30 to FIM 35	Contain public expenditure
1992	Fixed deductible for drugs in BRC increased from FIM 35 to FIM 45 and the reimbursement rate reduced from 50 to 40%. Reimbursement rate for drugs in LSR reduced from 90 to 80%. Some OTC products removed from the list of reimbursable medicines	Contain public expenditure
1993	Generic substitution operational	Contain public expenditure
1994	A fixed deductible, payable per purchase, introduced into SRCs. Fixed deductible for drugs in BRC increased from FIM 45 to FIM 50, and the reimbursement rate from 40 to 50%. An additional refund became payable only after the annual ceiling set to the patient's copayments exceeded by FIM 100	Contain public expenditure
	Direct price monitoring abolished	EEA agreement
1995	The Pharmaceutical Pricing Board became responsible for setting the wholesale price, in which the reimbursement is based	To control prices after EEA agreement
	Turnover tax for pharmaceuticals was replaced by a 12% VAT charge. The change increased retail prices about 7%	Increase general revenues for the government
1996	Generic substitution was replaced by generic prescribing	
1998	The criteria for new drugs to become eligible for Special Refunds reviewed. VAT was reduced from 12% to 8%. Compulsory stockpiling surcharge abolished. Pharmacists margin made more degressive. Wholesale prices set only for a fixed term. Wholesale prices reviewed and rationalized (1998–1999). Applications for reimbursable wholesale prices concerning products containing new active ingredient and, when necessary, those concerning other pharmaceutical products, shall include a health economic evaluation	Contain public expenditure
	ROHTO programme launched	To guide physicians towards more rational prescribing
1999	Subgroup of significant and expensive drugs introduced. Drugs in this group are reimbursed only if the illness fulfils certain criteria	Contain public expenditure

Table 4.3 (*Continued*)

Year	Reform/Change	Aim of the reform/Change
2003	Generic substitution introduced on 1st April. The prescribed medicinal product is substituted in a pharmacy by the cheapest, or close to the cheapest, generic alternative. Both, the prescribing physician and the purchasing individual have the power to refuse the substitution. The reimbursement payment will be based on the price of the dispensed product. The price of a medicinal product is considered to be close to the cheapest, when the price difference to the cheapest substitutable medicinal product, costing less than €40, is less than €2 or when the price difference to the cheapest substitutable medicinal product, costing €40 or more, is less than €3	Cut down pharmaceutical expenditure by introducing competition between pharmaceutical companies
2003–2004	Wholesale prices reviewed by the Pharmaceutical Pricing Board	
2004	The decisions relating to choosing the medical product eligible for reimbursements under SRC were given to the Pharmaceutical Pricing Board. The medicinal product must be in the BRC for at least 2 years before it can be changed to eligible for SPR	Contain public expenditure
2006	Change in calculation of reimbursements. Fixed deductibles for BSR were abolished and 42% reimbursement is calculated separately for each of the medical products. The former fixed deductible, payable per purchase for HSR were changed to a fixed deductible, payable per medical product. The former fixed deductible, payable per purchase for LSR were abolished and 72% of the cost of medical products belonging to the category are reimbursed	To simplify the system without reducing the level reimbursements payments
	Reduction of wholesale prices for all reimbursed medicines by 5%	Contain expenditure
	Both reimbursability and 'reasonable' are needed to be approved by Pharmaceutical Pricing Board	

BRC = Basic Refund Category; FIM = Finnish Mark, national currency before the Euro; HSR = Higher Special Refund Category; LSR = Lower Special Refund; OTC = Over the Counter; SRC = Special Refund Category; EEA = European Economic Area agreement; VAT = Value Added Tax. Source: [36, 49] and the author.

the original wholesale price was accepted. The same applies, if a pharmaceutical product containing the same effective ingredient or combination of medicinal substances were available at a considerably lower price, or if a particular product were available at much lower price in the Scandinavian countries or in the other EU member's states. In addition, since 1998 wholesale applications concerning pharmaceutical product containing a new active ingredient and, when necessary, those concerning other pharmaceutical products, shall contain a health economic report. After 1998 all decisions considering wholesale prices were defined for a fixed period, the maximum being five years. The Pricing Board has revived all wholesale prices for all prescribed medicines twice (during the years 1998–1999 and 2003–2005).

The latest revision of drug reimbursement system (in 2006) included changes in the calculation of reimbursements. The revision was designed in a way that it will not change the average share paid by NHI and patients. In the same year, all wholesale prices were reduced by 5% (this applied to the approved wholesale price). In the case that a producer had already sold the product below the approved price, the reduction was smaller or even nil. In addition, other measures were also tightened. The criteria for reimbursement for the subgroup of 'significant and expensive drugs' (introduced in 1999) were restricted. In 2006 this group was renamed 'Medicinal products Eligible for Restricted Basic Fund'. For example, the reimbursements of some expensive statin products were restricted to specific high-risk patient groups. Since the start of 2006, the reimbursibilty (i.e. clinical efficacy) of a medicinal product and its 'reasonable wholesale price' both needed to be approved by the Pharmaceuticals Pricing Board. Up until the end of 2005, a medicinal product with an approved 'reasonable wholesale price' became automatically reimbursable under the Basic Refund Category.

4.4.2.2 Generic Substitution

Voluntary generic substitution was introduced in 1993, whereby a medicine prescribed by a doctor was substituted in a pharmacy with a cheaper equivalent. The NAM approved a list at the outset of about 20 pharmaceutical agents that could be substituted, and both the doctor and the patient had to agree to the substitution. As the substitution was made on a cheaper equivalent it reduced both the reimbursements and the patients' expenses. However, generic substitution generated little interest, with about only 10 to 20 prescriptions per month throughout Finland [37].

Generic substitution was abandoned in 1996 and prescriptions using generic names were introduced. The idea was that the doctor will prescribe only by using a generic name, not the brand name, and the pharmacy will supply the patient with the least expensive product chosen from the available pharmaceutical equivalents. The amount of generic prescribing remained insignificant. Out of 2.6 million monthly prescriptions, on average only 600 were prescribed using a generic name [37].

Compulsory generic substitution became effective on 1st April 2003. The prescribed medical product is substituted in a pharmacy by the cheapest, or close to the cheapest, generic alternative. Both, the prescribing physician and the purchasing individual have the power to refuse a substitution. The reimbursement payment will be based on the price of the dispensed product. The price of a medicinal product is considered to be close to the cheapest, when the price difference to the cheapest substitutable medicinal product, costing less than €40, is less than €2 or when the price difference to the cheapest substitutable medicinal product, costing €40 or more, is less than €3.

4.4.2.3 Other Measures

The Government has also encouraged the development of advice for doctors on rational prescribing behavior, which is independent of (and will counterbalance) marketing by the pharmaceutical industry. Specific guidelines related to medicines do not exist, but ROHTO (a national educational programme for rational drug therapy and prescribing) was implemented during 1998–2002. By utilizing the available data on the usefulness and overall financial implications of various drug treatments, the program was aimed at improving prescription practices. Various approaches, including the continuing of evidence-based medical education, information delivery and providing prescribing feedback, were used simultaneously to encourage physicians to critically review their own prescribing practices [38, 39]. The ROHTO program was particularly designed to reach primary health care – that is, doctors working at health centers and those involved in occupational health care. In the wake of the perceived success of the ROHTO program, a new Pharmacotherapy Development Center was set up in 2003 with a remit to provide doctors with "... balanced information on new medicines and treatments".

In addition, attempts to affect the doctors' prescribing behavior by feedback information have been made. Since 1997, all doctors who have written over 200 reimbursed prescriptions during a calendar year have received a summary of their prescriptions and their costs from the Social Insurance Institution [40]. The data provided include the number of prescriptions and their distribution by patients' age and gender; total costs of the medications; average costs of a prescription; groups of medicines which have been most commonly prescribed by the doctor; and major groups which have caused the greatest costs. All relevant figures are compared with the average of the specialty of the doctor concerned. In addition, the Social Insurance Institution follows prescriptions at an individual level and, in serious cases, can also contact a doctor who prescribed the drugs.

4.4.3
Evaluation of Strategies

All of the above-mentioned measures have attempted to contain cost, and some also have tried to encourage rational prescribing in order to increase the cost-effectiveness of the system. However, since the measurement of outcomes (in terms of health gains) of change in the patterns of use of medicines is very difficult to evaluate, these changes will be analyzed only by using expenditure and price information.

4.4.3.1 Total Expenditure on Drugs

In spite of various measures, the growth in drug expenditure has been high. For example, during the years 1990–2005 the share of prescribed medicines (either partly

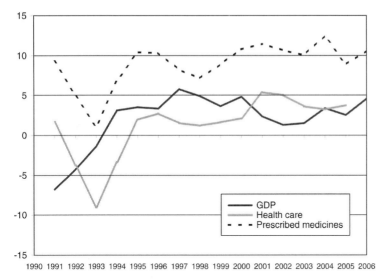

Figure 4.6 Annual changes (%) of GDP, health care expenditure
and expenditure on prescribed (under the NHI scheme)
medicines per capita between 1990 and 2006.

or totally reimbursed by NHI) of total health expenditure has increased from 7 to
13.5%. The share decreased only in 2005 (from 13.7 to 13.5%). During the same time
interval its share of GDP increased from 0.5 to 1.0%, but decreased somewhat only in
2006. The annual increase in the volume of expenditure on prescribed medicines per
capita (i.e. expenditure at constant prices[12]) has outpaced that of both GDP and total
health expenditure every year since 1990 (Figure 4.6). Expenditure on prescribed
medicines grew even during the severe economic recession of the early 1990s, while
first the GDP and later health expenditure fell by over 5% in one year alone. During
the 2000s the volume of prescribed medicines has increased each year by about 10%.

Judging by an international comparison of the pharmaceutical expenditure share
of GDP, Finland looks still to be a relatively low-spending country with a typical
growth path for such spending. However, judging by the pharmaceutical expenditure
share of total health expenditure, Finnish spending has been following a rising trend.
During the late 1990s and early 2000s the rate of increase of the share was faster than
in either Nordic countries or in other OECD countries [5, 41].

Pharmaceutical expenditure is typically private expenditure, which is driven by
demand. Total expenditure is determined by the number of patients using medicines,
the expenditure of medicines per patient, and the unit price of medicines. In
Figure 4.7 the annual changes in expenditure of prescribed medicines is divided

12) Expenditure at constant prices was calculated
using the retail price index of medicines. This
index does not take into account the qualitative change of medicines – that is, the fact
that newer and more expensive medicines
have been taken into use. Thus, the volume
change includes these qualitative changes
(see also footnote 4).

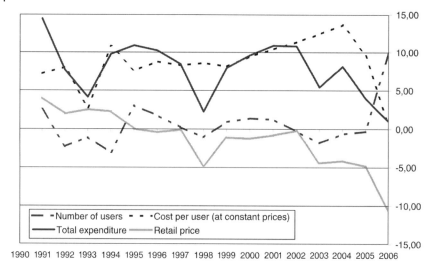

Figure 4.7 Annual changes (%) in factors determining expenditure of prescribed medicines reimbursed by the NHI in 1990–2006. The data show changes in price level, total number of users and volume per user (expenditure at constant prices).

into these three determinants. As can be seen, the main driver of cost escalation is the fact that volume/user (cost at constant prices) has increased annually by about 8% on average.

The only exception of the general trend is the year 2006, when total expenditure increased by only 1%. The exceptional figures for the year are due to changes introduced to the reimbursement system and to a 5% reduction of all approved wholesale prices at the start of the year. For example, the changes introduced to the reimbursement system significantly increased the actual number of users, since previously the users who purchased medicines that cost less than the fixed deductible payable by the patient were totally excluded from the reimbursements scheme (Table 4.3). However, as the data in Figure 4.6 indicate, per-capita expenditure at constant prices was still increased by about 10%. Thus, the price reduction (in total over 10%) is the main explanation for the low increase in total expenditure in 2006.

An analysis of regional development of pharmaceutical expenditure/per capita indicates [42] that this trend in volume of utilization has occurred in a similar way in all regions, while demand factors such as regional demographic, socioeconomic and needs factors explain very little of the increase in pharmaceutical expenditure. Although the factors behind the increase in the volume per user are not known exactly, it seems that it is due both to increasing the number of medicines used per person and the change in quality of utilization in terms of new and medicines that are more expensive. The regional variation in per-capita expenditure on prescribed medicines is also much lower than that of municipally provided services. This variation has even decreased since 1993 [30].

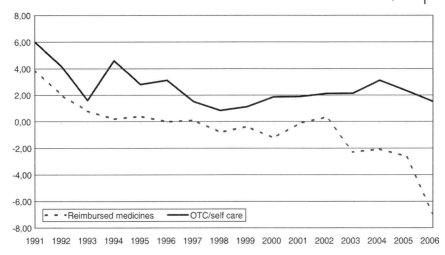

Figure 4.8 Annual changes (%) in wholesale prices of reimbursed
and OTC/self care products over the period 1990–2006.

4.4.3.2 Development of Prices

Most regulatory measures have attempted to contain whole prices. The regulatory
policy appears to have been rather successful, since the annual change in wholesale
prices decreased and, after 1997, the price level has even decreased absolutely. The
price regulation of over-the-counter (OTC) medicines was canceled in 1994, and since
then prices have been determined in the markets. As can be seen from Figure 4.8,
OTC wholesale prices have increased more rapidly than prices of prescribed
medicines, even at a time of price regulation, although the difference has been
increased during the 2000s.

The regulation of wholesale prices in Finland seems to be effective from a broader
perspective. According to an international comparison made in 2005 by IMS Health
(an international consulting and data services company), Finland is the second-
lowest of all European countries in terms of wholesale prices for the top 100 best-
selling pharmaceuticals. On average, the Finnish prices were about 11% lower than
the European average. Similar findings have been reported for smaller groups of
drugs across seven European countries [43]. On the other hand, the comparative
figures for retail prices give a different result, with Finland being ranked sixth
highest.

4.4.3.3 The Effect of Generic Substitution

The introduction of generic substitution in 2003 further decreased the price level of
reimbursed medicines, the decrease being greatest in the Basic Refund Category and
in the Lower Special Refund Category (Figure 4.9). On the other hand, the introduction of competition has not affected prices in the Higher Refund Category. It has been
estimated that generic substitution generated €88 million in cost savings during the
first 12 months (i.e. between 1st April 2003 to 31st March 2004). The patient's share of
savings was €39 million, whilst €49 million was saved in drug reimbursement

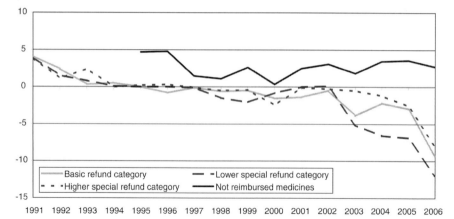

Figure 4.9 Annual changes (%) in wholesale prices of reimbursed medicines (basic, lower and higher refund categories) and nonreimbursed prescribed medicines between 1990 and 2006.

payments. About two-thirds of the savings were estimated to be due to reduction in price level as a result of increased competition, and one-third to product substitution [44]. During the first 12 months the savings due to substitutions were €20 million, in 2004 €31 million, in 2005 €26 million, and in 2006 €36 million; hence, the total was about €113 million. The effect of reduction due to increased price competition was not possible to evaluate for later years, since measures other than competition have caused prices to decrease (Table 4.3). The growth rate of expenditure on prescribed medicines slowed to 6% during the first year of generic substitution, but in 2005 the rate increased again to 8%. Thus, the impact of generic substitution may have been temporary.

4.5
Lessons from Finland

At the macro level, the Finnish systems seem to curtail the costs of health care rather well compared to other countries, with health expenditure per capita and as a percentage of GDP being lower than in many countries. This is in part due to the relatively low wages of health care personnel in Finland. In addition, cost control by municipalities (financed mainly by local taxes) has so far been very effective. Thus, the control of health expenditure might be achieved by a decentralization of the financial responsibility of health care to directly elected local governments which are responsible also for other public services and must cover the expenses, mainly by local taxes, the level of which they can decide themselves.

However, not much can be concluded about whether the Finnish system has been effective in terms of health gains because of difficulties in comparing outcomes [45]. In basic and secondary education the Finnish system has been indicated to perform internationally very well, with high technical and cost efficiencies [46, 47]. This evidence cannot – as such – be generalized into health care, since in basic and secondary education the regulation of central government is stronger than in health care, and state subsidies are also partly earmarked based on the number of pupils and estimated unit costs.

Although the cost control at macro level seems to work well in Finnish health care, there are wide regional differences in access, utilization, outcomes and efficiency between municipalities and hospital districts. So far, the general Government has tried to affect these by 'softer methods' such as providing nonbidding recommendations and supporting municipalities and hospital districts with funding to develop projects. Powerful regulatory actions have not been used, except in the introduction of legislation for waiting-time guarantees, in which case the regulation has been indicated to have positive effects [48].

In practice, the local projects and experiments have emerged in quite different directions, thus increasing the diversity in models of the production of health services within the country. In order to assess the effect of diversity, to learn from best practice and to provide government tools for directing health policy, more attention should be paid to developing performance indicators. Proper outcome measurement requires the coordinated, long-term development of methods and data collection, and this has not been a priority for the Ministry of Social Affairs and Health. The indicators have been developed on voluntary basis, initiated by researchers and afterwards implemented together with producers (hospital districts) by using financial support from research funds.

A recent trend in Nordic health care systems has been that of *centralization*. The reforms implemented or prepared in the 21st century (in Norway 2001, Denmark and Finland 2007, Sweden suggested by the year 2015) have been focused on the production side, mainly by centralizing the decision making via decreasing the number of local governments/authorities responsible for providing services. However, the Finnish reform (as opposed to other Nordic reforms) again indicates the Government's unwillingness or incapacity to perform a real reform, and much hope has been placed on the voluntary actions of municipalities to join together in order to improve the efficiency of production.

The relatively low wage level of health sector employees in Finland can mainly be explained via the centralized collective wage agreement system, in which it is has been very difficult to increase the relative wages of specific employee groups. On average, Finnish doctors and nurses have resorted to strike action once every 10 years, although in the long run their wages have followed the general wage level. The most recent (2007) contract between the Union of Health and Social Care Professionals (representing nurses) and municipal employer organizations introduced 'productivity' into wage determinations. After 2011, the wage level of nurses will depend on the growth of the total number of persons working in health and social services, such that a decrease (or at least not an increase) in the total number of employees between the years 2006 and 2010 will lead to an increase in their wage levels.

Cost containment in medicines has been based mainly on price regulation and, recently, to some extent on increasing the role of price competition in a regulatory framework. These measures have curtailed quite effectively the price development, and the price of existing old products has decreased – indicating that prices have been rather loose. A new suggestion is to introduce a generic reference price model at the start of 2009 [49]. However, until now the effects of various measures have been rather temporary, and the main driver of expenditure has been the use of new and more expensive drugs.

The public financing of the two funding systems has been independent of each other, and not integrated. For example, municipalities (public hospitals and nursing homes) pay for the drug expenditures of inpatient care, whilst for outpatient care both the patients and the NHI contribute to the expenditure. The budget for the NHI is open-ended, while the municipalities' budgets are constrained. Thus, in the present system the municipalities have incentives to find those care alternatives which shift financing to others (e.g. drugs and other services covered by the NHI), although the alternative will be against patients, needs, total cost or the effectiveness of services. On the other hand, many new and expensive outpatient drugs (e.g. in cancer care) are not covered by the NHI, but rather are provided by hospitals and thus financed by municipalities. This creates an unnecessary use of hospital care by patients in order to obtain medication. Moreover, the development of new care models which substitute hospital or nursing home care, and are provided by the private sector, emphasize the significant perverse economic incentives associated with a two-tier financing system. For example, the OEDC has suggested transferring the responsibility for public spending on medicines from the Social Insurance Institution to municipalities. According to interviews with Finnish health care decision makers, the existence of the two-tier funding system has now been widely recognized as being one of most important problems in Finnish Health care (M. Pekurinen *et al.*, unpublished results).

References

1 Häkkinen, U. (1995) Health care in Finland: current issues, in *The Nordic Lights. New Initiatives in Health Care Systems* (eds A. Alban and T. Christiansen), Odense University Press, Odense, pp. 141–148.

2 Mossialos, E. and Le Grand, J. (eds) (1999) *Health Care and Cost Containment in the European Union*, Ashgate.

3 Ministry of Social Affairs and Health (1993) Health for All by The Year 2000. Revised strategy for co-operation.

4 Ministry of the Interior (2005) Decision to Investigate A Project to Restructure Municipalities and Services, 11 May 2005 SM043.

5 OECD (2005) *OECD Review of Health Systems Finland*, OECD, Paris.

6 Häkkinen, U. and Lehto, J. (2005) Reform, change and continuity in Finnish health care. *Journal of Health Politics, Policy and Law*, 30 (1–2), 76–96.

7 Häkkinen, U. (2005) The Impact of changes in Finland's health care system. *Health Economics*, S1, s101–18.

8 Järvelin, J. (2002) Health Care Systems in Transition: Finland. European Observatory on Health Care Systems.

9 Vuorenkoski, L. (2008) Health System in Transition: Finland European Observatory. Health Care Systems.

10 STAKES (2007) Health Care Expenditure and Financing in 2005. Tilastotiedote 2/2007, 26/1/2007.

11 OECD (2000) System of Health Accounts, OECD, Paris.

12 Finnish Government (1992) Government Decision in Principles on Actions to Increase Balance in Public Finance 14/10/1992.

13 Finnish Government (1992) Government Bill (291/1992) for user charges in social and health care. http://www.finlex.fi/f§i/esitykset/he/1992/19920291.

14 Oates, W. (1999) An essay on fiscal federalism. *Journal of Economic Literature*, **37**, 1120–1149.

15 Häkkinen, U., Linna, M. and Salonen, M. (1994) Korvausmenettelyn ja kuntakoon vaikutus erikoissairaanhoidon taloudelliseen riskiin (The effects of reimbursement methods and size of municipality on economic risk of specialised care). *Suomen Lääkärilehti (Finnish Medical Journal)*, **49**, 2454–2458.

16 Mikkola, H. (2002) Empirical studies on Finnish hospital pricing methods, *Acta Universitatis Oeconomicae Helsinginesis*, A-203.

17 Häkkinen, U. and Linna, M. (2005) DRGs in Finnish health care. *Euro Observer*, **7** (4), 5–8.

18 Ihalainen, R. (2007) *Sopimusohjaus Erikoissairaanhoidon Palvelujen Tuottamisen Ohjauksessa (Steering by Contracts in Service Provision of Specialised Medical Care)*, 1234th edn, Tampere Universitatis Press, Tampere.

19 Ministry of Social Affairs and Health (2002) Decision in Principle by the Council of State on securing the future of health care. Helsinki, Internet: http://www.stm.fi/english/eho/publicat/bro02_6/bro02_6.pdf (17 January 2003).

20 Eskola, J., Höckerstedt, K., Mäkäräinen, H., Oxman, A. and Lampe, K. (2004) The future of FinOHTA an External Review. Stakes, FinOHTA report 23.

21 Lehto, J. and Blomster, P. (2000) Talouskriisin jäljet sosiaali- ja terveyspalvelujärjestelmässä (The marks of economic recession in health and social care system). *Kunnallistieteen Aikauskirja*, **28**, 44–60.

22 Häkkinen, U. and Järvelin, J. (2004) Developing the formula for state subsidies for health care in Finland. *Scandinavian Journal of Public Health*, **32**, 30–39.

23 Luoma, K. (2000) Terveyskeskusten tuottavuus ja panosten käytön tehokkuus 1990-luvulla (Productivity changes and efficiency in Finnish health centres in the 1990s). *Sosiaalilääketieteellinen Aikakauslehti (Journal of Social Medicine)*, **37** (3), 207–215.

24 Linna, M. (1999) *Measuring Hospital Performance: The Productivity, Efficiency and Costs of Teaching and Research in Finnish Hospitals*, Stakes Research Reports 98, Gummerus Printing, Jyväskylä.

25 Järviö, M- and Räty, T. (2003) Terveyskeskukset 1997–2001 (Health centers 1997–2001), in *Kunnalliset Palvelut. Terveyden ja Vanhustenhuollon Tuottavuus (Municipal Services, Productivity in Health Care and Services for The Elderly)* (eds R. Hjerppe, A. Kangasharju and R. Vuorento), Government Institute for Economic Research, Helsinki, pp. 25–36.

26 Linna, M. and Häkkinen, U. (2004) Erikoissairaanhoidon Tuottavuuden Kehitys 1998–2002 (Productivity in specialised care 1998–2002), in *Sairaaloiden tuottavuus. Benchmarking-tietojen käyttö erikoissairaanhoidon toiminnan suunnittelussa, seurannassa ja arvioinnissa (Use of Benchmarking Data in Planning, Monitoring and Evaluating Specialised Care)* (ed. M. Junnila), Stakes raportteja, 280, Helsinki, pp. 44–71.

27 Luoma, K. and Moisio, A. (2005) *Kuntakoko, kuntien menot ja palvelujen tuotannon tehokkuuserot (Size of Municipality, Municipal Expenditure and Efficiency Differences)*, Government Institute for Economic Research Muistiota 69.

28 Luoma, K., Moisio, A. and Aaltonen, J. (2007) Secessions of Municipal Health Centre Federations: Expenditure and Productivity Effects, Discussion paper 425, Government Institute for Economic Research, Keskustelualoitteita.

29 Nolte, E. and McKee, M. (2004) *Does Health Care Save Lives? Avoidable Mortality Revisited*, The Nuffield Trust, London.

30 Hujanen, T., Pekurinen, M. and Häkkinen, U. (2006) Terveydenhuollon ja Vanhustenhuollon Alueellinen Tarve ja Menot 1993–2004(The Regional Need and Expenditure of Health Care and Care of The Elderly 1993–2004), Stakes, Työpapereita, 11/2006.

31 Lehtonen, L., Andersson, S., Hallman, M., Lavonius, M., Leipälä, J., Tammela, O. *et al.* (2007) PERFECT- Keskoset. Hyvin ennenaikaisten keskosten hoito, kustannukset ja vaikuttavuus. (Perfect low-birth-weight infants. Care, cost and outcomes of very preterm infants), Stakes, Työpapereita, 16/2007.

32 Häkkinen, U., Idänpään - Heikkilä, U., Keskimäki, I., Klaukka, T., Peltola, M., Rauhala, A. *et al.* (2007) Perfect-Sydäninfarkti. Sydäninfarktin hoito, Kustannukset ja Vaikuttavuus. (Perfect AMI, The Care, Cost and Outcomes of Acute Myocardial Infarction Patients), Stakes, Työpapereita, 15/2007.

33 Meretoja, A., Roine, R.O., Erilä, T., Hillbom, M., Kaste, M., Lönnqvist, J. *et al.* (2007) Perfect- Stroke. Hoitoketjujen Toimivuus, Vaikuttavuus ja Kustannukset Aivoverenkiertohäiriöpotilailla. (Perfect-Stroke, the Performance, Effectiveness and Cost of Treatment Episodes among Stroke Patients), Stakes, Työpapereita, 23/2007.

34 Remes, V., Peltola, M., Häkkinen, U., Kröger, H., Leppilahti, J., Linna, M. *et al.* (2007) Perfect- Tekonivelkirurgia. Lonkan ja Polven Tekonivelkirurgian Kustannukset ja Vaikuttavuus. (Perfect-Endoprosthesis Surgery. The Cost and Outcomes of Hip and Knee Replacements), Stakes, Työpapereita, 29/2007.

35 Rautava, L., Lehtonen, L., Peltola, M., Korvenranta, E., Korvenranta, H., Linna, M. *et al.* (2007) The Effect of birth in secondary- or tertiary-level hospitals in Finland on mortality in very preterm infants: a birth-register study. *Pediatrics*, 119 (1), e257–e263.

36 Pekurinen, M. and Häkkinen, U. (2005) Regulating pharmaceutical markets in Finland. Stakes Discussion Papers 4/2005.

37 Martikainen, J., Rajaniemi, S. and Klaukka, T. (1999) Reimbursement of Medicine Cost in 1990s. Background and Development. The Social Insurance Institution Social Security and Health Research. Working papers 11/1999.

38 Helin-Salmivaara, A., Huupponen, R., Klaukka, T. and Horppu, K. (2002) Focusing on changing clinical practice to enhance rational prescribing – collaboration and networking enable comprehensive approaches. *Health Policy*, 66, 1–10.

39 Nikkarinen, T., Huvinen, S. and Brommels, M. (2002) National Consensus and Local Consideration. Changing the Prescription Practices by Means of Training, an Evaluation Report of the ROHTO Project. Reports of the Ministry of Social Affairs and Health 1.2002.

40 Klaukka, T. (2002) The Finnish data base on drug information. *Norwegian Journal of Epidemiology*, 11 (1), 19–22.

41 OECD (2007) OECD Health Data, 2007.

42 Mikkola, H., Häkkinen, U. and Klaukka, T. (2002) The Impact of Regional Factors on the Increase in Medicine Cost. Paper presented at the Nordic Health Economists' Meeting, Helsinki, August.

43 Martikainen, J., Kivi, I. and Linnosmaa, I. (2005) European prices of newly launched reimbursable pharmaceuticals – a pilot study. *Health Policy*, 74, 235–246.

44 Paldan, M. and Martikainen, J. (2005) Lääkevaihdon ensimmäinen vuosi tilastoina (The first year of generic substitution in statistics, in *Lääkevaihdon Ensimmäinen Vuosi (The First Year of Generic Substitution*, (eds R. Ahonen and J. Martikainen), 68, KELA Sosiaali-ja terveysturvan katsauksia, Helsinki, pp. 27–38.

45 Häkkinen, U. and Joumard, I. (2007) Cross-Country Analysis of Efficiency in

OECD Health Care Sector: Options for Research. OECD Economic Department Working Paper No. 554; 2007.

46 Sutherland, D., Price, R., Joumard, I. and Nicq, C. (2006) Performance Indicators for Public Spending Efficiency in Primary and Secondary Education. OECD Economics Department Working Paper No. 546; 2006.

47 OECD (2007) PISA™ 2006 Science Competencies for Tomorrow's World. Volume 1: Analysis; 2007.

48 Pekurinen, M., Mikkola, H. and Tuominen, U. (eds) (2008) Hoitotakuun Talous- Hoitotakuun Vaikutus Terveydenhuollon Menoihin, Toimintaan ja Sairausvakuutuskorvauksiin. (The Economics of Waiting Time Guarantee – the Effects on Waiting Time Guarantee on Health Expenditure, Performance and NHI Reimbursements), Stakes, Raportteja, 5/2008.

49 Ministry of Social Affairs and Health (2007) Memorandum of the Working Group on the Introduction of a Reference Price System for Medicines. Reports of the Ministry of Social Affairs and Health, 46; 2007.

5
Germany

Markus Wörz and Reinhard Busse[1]

Population (million)	82.5
GDP per capita (US$ PPP)	29 800
Health spending as % of GDP	10.7
Public health spending as % of total spending	76.9
Health spending per capita (US$ PPP)	3287
Acute care beds per 1000 population	6.4
Practicing physicians per 1000 population	3.4
Life expectancy at birth (years)	79
Infant mortality per 1000 live births	3.9

Strategies used

1. Global budgets and spending caps
2. Case-based (DRG) fee system (fully effective 2010)
3. Cost shifts to private households
4. Free choice of sickness fund
5. Enhanced role of joint self government
   (association of sickness funds and providers)

5.1
A Review of the Major Structural and Operational Characteristics of the German Health Care System

Germany's health care system is the archetype of a Bismarckian health care system, its major structural component being the Statutory Health Insurance (SHI). As Figure 5.1 shows, about 90% of the population are insured there, whereas most of the remaining 10% are insured with private health insurance. Therefore, this chapter

1) We are very grateful to Miriam Blümel for able
   and diligent research assistance.

Cost Containment and Efficiency in National Health Systems: A Global Comparison
Edited by John Rapoport, Philip Jacobs, and Egon Jonsson
Copyright © 2009 WILEY-VCH Verlag GmbH & Co. KGaA, Weinheim
ISBN: 978-3-527-32110-0

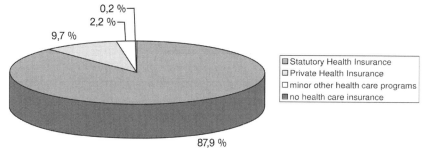

0,2 %
2,2 %
9,7 %

Statutory Health Insurance
Private Health Insurance
minor other health care programs
no health care insurance

87,9 %

Figure 5.1 Different types of health care coverage in 2003 (in %).[2] Note: Minor health care programs refer to beneficiaries of public assistance, free health care for the police and the Federal Armed Forces and health care for war pensioners.[3] ([3], 730)

focuses primarily on the strategies of containing costs and improving efficiency within the SHI.

The Federal Ministry of Health is primarily responsible for the institutional and legal design of statutory health and long-term care insurance, as well as their supervision.[4] The usual policy process is that the Federal Ministry of Health drafts reform bills which are then brought to Federal Parliament where they are debated, modified and passed.

The Social Code Book V is the regulatory framework of the SHI system. There, the goals of the SHI are defined, along with the rules that guide it. Stated goals and the fundamental principles of the SHI can be described as follows (the following is in parts directly quoted from Ref. [4]: S133f.):

- The SHI has the task to maintain, restore or improve the health of the insured.
- The insured must take responsibility for their health.
- The provision of services and their financing must be based on the principle of solidarity, which has the following features:
  – contributions proportional to income up to an income threshold and free insurance of family members who are not gainfully employed;
  – no differentiation of contribution rates according to age, gender or health risk;
  – provision of medically necessary services according to the principle of appropriateness in a sufficient and efficient way.

- Sickness funds and service providers have to guarantee a needs-based and consistent provision of services which meets the standards of medical knowledge. Health care for the insured must be sufficient and expedient, must not be in excess

---

2) Data on health care coverage for the year 2007 will be available in April 2008.

3) War pensioners represent a very small, almost negligible, minority. In 2003 there were 19 000 war pensioners ([2]: 753).

4) Other competences of the Federal Ministry concern pharmaceuticals, health protection, prevention and biotechnology (cf. for more on these competences: Ref. [1]: 30ff.).

of what is medically necessary, and must be provided efficiently, humanely, and up to the required quality.

- Sickness funds and service providers have to provide services in such a way that the contribution rates for the SHI remain stable (the principle of contribution rate stability). As a basic rule, this shall be guaranteed by limiting the increases of reimbursements for health care providers to the increases in the contributory income of SHI insurees.

- The service guarantee for the medical outpatient treatment lies with the Federal Association of SHI Physicians and the regional physicians' associations, while the service guarantee for the inpatient medical treatment lies with the Federal States.

5.1.1
Key Aspects of Sickness Funds and Providers of Health Care

The statutory sickness funds for health care are legally divided into seven groups. On 1 January 2007 there were 242 sickness funds in the following seven groups:

- 16 general regional funds (AOKs)
- 189 company-based funds (BKKs)
- 16 guild funds (IKKs)
- 10 'substitute' funds (divided into 'white-collar' and 'blue-collar')
- 9 farmers' funds
- 1 miners' fund and 1 sailors' fund ([5], 37)

The seven groups are organized into associations both on the Federal and on the Federal State (Bundesland) level. Among others, these associations serve as partners within joint self-government (for the tasks of joint self-government, see below). Whereas the benefit catalogue that individual sickness funds have to offer is rather highly regulated and more or less the same for all sickness funds, they are free to set the contribution rates for insurees.[5] Contribution rate differentials became smaller during the 1990s (see Section 5.2.3 for more information on this).

Table 5.1, which provides an overview on the structures of providers in the ambulatory and hospital care sector, shows that the majority of physicians work in solo practices. In contrast to many other health care systems, there is a very high number of specialists in the outpatient sector in Germany (in the year 2005 there were approximately as many general practitioners (GPs) as there were specialists in the ambulatory sector). There are no data available on the number of group practices, or their average size. A comparatively new organizational form is that

---

5) As of 2009, a uniform contribution rate will be set centrally by the government; moreover, the associations of the seven groups of sickness funds will be merged into one central association (see Section 5.3 for more details).

Table 5.1 Overview on providers in the ambulatory and inpatient sector (2005).

Ambulatory sector		Inpatient sector	
Provider	**Active physicians (in 1000)**	**Provider**	**No. of hospitals**
Solo practices	82.5	General acute	1846
Group practices	43.0	Psychiatric/psychotherapeutic	234
Medical care centers (overall: 491)	1.9	Pure day and night clinics	59

Source: Refs [6, 7] and own calculations; absolute number of group practices are not available (the number of solo practices should approximately correspond to the number of working physicians in solo practices.

of medical care centers, which are multidisciplinary, physician-led institutions for ambulatory care services. These were introduced in 2004, and so it can be expected that this number will rise in the future. Whereas, solo and group practices operate on a private for-profit basis, medical care centers can be both for-profit and nonprofit.

In order to acquire political representation and self-government, ambulatory physicians must be members of the Regional Associations of SHI Physicians. The 17 Regional Associations of SHI Physicians are unified in the Federal Association of SHI Physicians, which is their peak association. Hospitals are organized in the German Hospital Federation (analogously to the sickness funds, both forms of organization are present on the regional (Land) and federal level).

A central feature in relation to the organization and management of the German health insurance system is the delegation of many tasks to the *joint self-government* between statutory sickness funds and providers of health care. Core elements of joint self-government are collective agreements between the associations of the sickness funds and the associations of SHI physicians. For example, the total remuneration for ambulatory care physicians (for total remuneration, see Section 5.2.1.2) is negotiated on the Land level within such collective agreements. There are several bodies in which joint self-government takes place, the most important being the Federal Joint Committee (*Gemeinsamer Bundesausschuss* – founded rather recently in 2004). Normally, within joint self-government statutory sickness funds act 'jointly and uniformly' – that is, there is hardly any discretion for sickness funds for individual action.

As will be described in more detail in Sections 5.2.3 and 5.2.4, there were two important – yet somehow contradictory – strategies for more efficiency and effectiveness of the health care system. On the one hand, the system of joint self-government has been gradually fleshed out and was equipped with more and more competencies. On the other hand, possibilities for direct agreements between individual sickness funds and individual providers or groups of health care providers beyond joint self-government were also extended.

5.1.2
Key Aspects on Financing of Health Care

As the data in Table 5.2 show, the total health expenditure as a proportion of GDP has increased rather moderately by 1.1% since 1992. However, the expenditure of SHI as a proportion of GDP remained virtually constant by a marginal increase of 0.1%. As will be discussed in more detail below, one important reason for this stability must be seen in the imposition of budgets on different sectors of expenditure of the SHI, in particular of pharmaceuticals, ambulatory care and hospitals.

The data in Table 5.3 show that private sources of health expenditure have risen considerably (i.e. by 4% since 1997). This is in particular so because of the increase of out-of-pocket payment (this point is discussed in Section 5.2.2).

5.2
Major Cost-Containment and Efficiency-Seeking Strategies

The major objective of health care reform in Germany has been – at least since the late 1970s until the early 2000s – the containment of costs ([1]: 186). As in other countries, cost containment has focused predominantly on controlling fees and overall spending, since imposing limits on providers is politically easier to do than imposing costs and restrictions directly on patients. The cuts of the first type are less visible to voters than, for example, higher copayments for patients ([10]: 703). It was only during the 1990s that other values – such as the effectiveness of technologies and cost-effectiveness or efficiency, and even more recently unclear concepts such as 'innovation' – have gained increased political and public attention.

In the following sections we present details of the important cost-containment and efficiency-seeking strategies introduced since the start of the 1990s. One suitable way of classifying cost-containment strategies is to distinguish between budget setting, budget shifting and direct and indirect control measures ([11]: 62ff.). In a similar vein, we distinguish between those strategies which aim directly at containing public or statutory health care costs and those which rather strive indirectly for the containment of costs. Budget setting and budget shifting are of the first type. We furthermore distinguish budget-setting techniques into global budgets and spending caps. Budget shifting strategies occur primarily in two ways: the increase of copayments; and the exclusion of statutory benefits. We also describe two strategies which aim indirectly to increase efficiency: (i) increased competition between sickness funds (and the concomitant strengthening of risk-adjustment mechanisms); and (ii) an increase of competition between providers on the one hand and step-by-step development of competencies for German joint self-government on the other hand. Somewhere in between these different strategies lie measures which concern the way in which medical services are reimbursed (e.g. per day or case fees for hospital services). This topic is dealt with in the next section, which provides details of global budgets and spending caps, as they are inherently related and affect each other.

Table 5.2 Main sources of finance, in percentage of GDP, 1992–2005.

	1992	1993	1994	1995	1996	1997	1998	1999	2000	2001	2002	2003	2004	2005
Total health expenditure	9.6	9.6	9.8	10.1	10.4	10.2	10.2	10.3	10.3	10.4	10.6	10.8	10.6	10.7
Expenditures on SHI	6.0	5.9	6.0	6.1	6.2	6.0	6.0	6.0	6.0	6.1	6.2	6.3	6.0	6.1
SHI expenditures on hospitals	2.1	2.2	2.3	2.3	2.3	2.3	2.3	2.2	2.2	2.2	2.2	2.2	2.2	2.3
SHI expenditures on physician's practices	1.1	1.1	1.1	1.1	1.2	1.1	1.1	1.1	1.1	1.1	1.1	1.1	1.1	1.1
Other SHI expenditures	2.8	2.5	2.6	2.7	2.8	2.6	2.6	2.7	2.7	2.8	2.8	2.9	2.7	2.7

Source: http://www.gbe-bund.de (accessed 23 August 2007).

Table 5.3 Main sources of finance, in percentage of total health expenditure, 1992–2005.

	1992	1993	1994	1995	1996	1997	1998	1999	2000	2001	2002	2003	2004	2005
Public sources	77.8	76.9	77.1	78.0	78.6	77.1	76.3	76.1	76.0	75.5	75.5	74.9	73.0	73.1
Taxes	11.2	11.7	11.3	10.7	9.3	7.2	6.7	6.5	6.4	6.4	6.3	6.2	6.1	5.7
Statutory health insurance	62.6	61.0	61.6	60.3	59.6	58.8	58.5	58.5	58.3	58.2	58.3	58.0	56.3	56.8
Statutory retirement insurance	2.2	2.4	2.3	2.3	2.3	1.6	1.5	1.5	1.6	1.6	1.6	1.5	1.5	1.5
Statutory accident insurance	1.8	1.9	1.9	1.8	1.8	1.8	1.7	1.7	1.7	1.7	1.7	1.7	1.7	1.7
Statutory long-term care insurance	—	—	—	2.8	5.6	7.7	7.9	7.9	7.9	7.6	7.6	7.5	7.5	7.5
Private sources	22.2	23.1	22.9	22.0	21.4	22.9	23.7	23.9	24.0	24.5	24.5	25.1	27.0	26.9
Out-of-pocket payments/NGOs	10.3	10.8	10.9	10.2	9.8	10.7	11.4	11.6	11.7	11.9	11.9	12.2	13.7	13.5
Private health insurance	7.4	7.8	7.7	7.7	7.5	8.0	8.0	8.2	8.3	8.4	8.5	8.7	9.0	9.2
Employer	4.4	4.5	4.3	4.2	4.1	4.2	4.2	4.1	4.1	4.2	4.1	4.2	4.2	4.2

NGO: non-governmental organization. Source: http://www.gbe-bund.de (accessed 23 August 2007) and own calculations.

5.2.1
Global Budgets and Spending Caps

It was argued that budgets ". . . may be a necessary but not a sufficient condition to ensure cost containment" ([12]: 49). Budgets are used in particular (though not only) in three areas of the German health care system, namely pharmaceuticals and the outpatient and inpatient sectors.

Global budgets (i.e. strictly set or target budgets) for: (i) a certain geographical area respectively a certain collective of providers in for example a Federal State or a region; or (ii) an individual provider for example a hospital, were – and still are – in force to contain costs.

Spending caps, defined as maximum expenditure limits that act in a similar manner to budgets, have been used primarily for prescribed pharmaceuticals, both on the level of physicians collectively and individually. (Note: Budgets and spending caps in Germany have always been defined on the provider or prescriber side, and not on the side of the payers – that is, the sickness funds).

A major break occurred in the German health care system in 1993 when the Health Care Structure Act (HCSA, *Gesundheitsstrukturgesetz* – GSG) came into force. This law contained many important reform measures, and its central component has been the reinforcement or introduction of budgets and spending caps for pharmaceuticals and for outpatient and inpatient care. Consequently, it essentially marks the beginning of an era of budgets of varying regulatory strictness. An overview of these different periods is provided in Table 5.4.

5.2.1.1 Pharmaceuticals

The development of 'drug budgets', which are *de facto* spending caps, can be divided into four stages [13]: in the first period (until 1992) there were no spending caps at all (for a more encompassing description of the regulation of pharmaceuticals before 1993, see Ref. [14]: 111ff.); in the second phase (1993–1997) there were collective spending caps; in the third phase (1998/99–2001) there were both collective spending caps as well as caps in the form of 'target volumes' for individual ambulatory physicians' practices; and in the fourth period (since 2002) there have been individual spending caps only. However, due to the introduction of drug procurement through sickness funds, these have become less important since 2007.

The HCSA introduced a collective spending cap for drugs prescribed in the ambulatory sector – that is, a limit to the total amount spent by statutory sickness funds on pharmaceuticals in the western part of Germany. The amount of the spending cap was set to the amount spent by sickness funds in 1991, reduced by savings assumed to have been realized by simultaneously passed policy measures ([14]: 114). In combination with these other measures, the spending of statutory sickness funds was 18.8% lower in 1993 than in the year before ([1]: 148). From 1994 to 1997 there were regional collective spending caps in the whole of Germany, though these were subject to negotiations between the associations of sickness funds and the

Table 5.4 Cost-containment through global budgets and spending caps, 1989–2007.

Note: The darker the background, the more strictly regulated the sector.

	Ambulatory care	Hospitals	Pharmaceuticals
1989–1992	Negotiated regional fixed budgets	Negotiated target budgets at hospital level	No spending caps
1993	Legally set regional fixed budgets	Legally set fixed budgets at hospital level	Legally set national spending cap
1994			
1995			Negotiated regional spending caps
1996	Negotiated regional fixed budgets	Negotiated target budgets at hospital level	
1997			Negotiated target volumes for individual practices
1998	(Target volumes for individual practice)[a]		
1999			Legally set regional spending caps
2000	Negotiated regional fixed budgets with legally set limit	Negotiated target budgets at hospital level with legally set limit	Negotiated regional spending caps[b]
2001			
2002			
2003	Legally set regional fixed budgets	Legally set target budgets at hospital level[c]	Negotiated target volumes for individual practices at regional level
2004	Negotiated regional fixed budgets with legally set limit	Negotiated target budgets at hospital level with legally set limit (since 2005: increasing activity-related component)	
2005			
2006			
2007			

Source: Ref. [1]: 164, slightly modified and updated.

[a]Legally, but not implemented (1997 status was kept).

[b]Due to the ministerial announcement of the spending cap lifting in January 2001, they were not enforced for 2001.

[c]Except for hospitals introducing DRGs already on a voluntary basis.

(then) 23 regional physician's associations[6] in Germany. As these spending caps were exceeded in some of the regions in 1994, 1995 and 1996, negotiations were conducted between sickness funds and physicians' associations regarding repayments of the latter, and it was agreed to even-out the overspend in the years to follow. Overall, these spending caps proved to be an effective method for the short-term reduction and long-term modification of pharmaceutical expenditure ([15]: 344).

The collective spending caps were formally abolished as of 1998 and replaced by individual prescription caps which set practice-specific target volumes. If ambulatory physicians exceeded 125% of these target volumes then they were required to compensate the respective sickness fund (apart from some exception rules, for example for certain high-cost drugs or some patient groups). In 1999, the more strict regional spending caps were reintroduced. The regional physicians' association became liable for any overspending, up to 105% of the budget; however, such collective liability claims were never executed as there were uncertainties in relation to the possibility of charging someone without individual infringement [13].

Pharmaceutical budgets on regional level were legally abolished at the end of 2001, and in 2002 the liability for negotiated target volumes for individual practices was legally fixed. As a consequence of this, statutory sickness funds were required to accept target volumes. In contrast to this, before 2002 the associations of sickness funds could insist on global budgets only as a means of budgetary regulation [13].

Another reform measure which is related to budgets and cost containment and efficiency is that of *reference pricing* of pharmaceuticals. This was introduced in 1989, and establishes reimbursement thresholds for certain classes of pharmaceuticals that are defined by the Federal Joint Committee. The impact of reference pricing on cost containment is limited. Although there is a clear incentive for pharmaceutical companies to set prices in line with reference prices, there are no incentives for any further price reductions [16]. It is difficult to determine the savings effect of the reference pricing system, as a counterfactual price development without reference prices cannot be simulated. Moreover, reference prices interact with other cost-containment policies [16, 17]. In retrospect, it is likely that spending caps had the most significant effect of all demand-side interventions in relation to pharmaceutical expenditure [15].

During recent years, health policy concerning pharmaceuticals has shifted its attention away from cost containment through spending caps, first towards a stricter evaluation of effectiveness (from 2004), then to rebates negotiated between sickness funds and drug producers based upon a procurement process (from 2007), and finally through an economic evaluation of cost-effectiveness (for which the methods are not yet fixed).

---

6) In 2007 there were 17 such regional physician's associations (one for every Federal State with the exception of North Rhine-Westphalia, where there are two regional physicians' associations).

5.2.1.2 **Ambulatory Care**

Almost all payment procedures for ambulatory physicians over the whole period reviewed are carried out within a two-stage process that takes place within joint self-government. First, the sickness funds pay a global budget, the total remuneration for medical services (Gesamtvergütung) to the regional associations of SHI physicians. The way in which this total remuneration is calculated and negotiated has varied over time. Second, the regional associations of SHI physicians distribute the payments to ambulatory care physicians according to varying criteria. In the following we briefly describe how the first and second stages of this process have evolved.

Development of Total Remuneration for Medical Services   In 1987, growth in expenditure on ambulatory physicians was limited to the growth in income per statutory sickness fund member ([18]: 9), but despite this measure physician expenditure continued to grow above the intended limits. When the HCSA came into force the total remuneration was set more strictly, with the law mandating that growth in physician expenditure between 1993 and 1995 be capped to the growth in income of statutory sickness fund members. As opposed to the situation before 1993, however, possibilities to exceed the budget were virtually nonexistent ([18]: 11). After 1995 the limitations for increases in total remuneration were somewhat relaxed, but re-tightened after the change in government in 1998. From 2009, the total remuneration will no longer be tightly budgeted as the morbidity of the insured must be included in its development.

Development of the Distribution of Total Remuneration for Medical Services to Ambulatory Physicians   In 1977, the so-called Uniform Value Scale (*Einheitlicher Bewertungsmaßstab* – EBM) was introduced for (West-) German physicians in the ambulatory sector. The EBM assigns 'point values' to be used by all sickness funds for the reimbursement of fees-for-services, and can be applied in two ways: (i) with fixed budgets as a relative value for distributing proportional payments out of the global budget; or (ii) with a predetermined conversion factor that serves as the basis for direct payment without fixed upper limits ([19]: 308). Since the total remuneration has been fixed in Germany since 1987 (with varying degrees of strictness), the first mode applies. The point values (the base rate or level of prices for medical services in the ambulatory sector) are determined by the regional associations of SHI physicians. The combination of both fixed upper limits and a fee-for-system leads to a sort of prisoner's dilemma situation: Independently of the behavior of other physicians, an individual physician can maximize his or her share of the total remuneration by expanding services so that the dominant strategy will be to expand services to patients ([4]: S139; [20]).

   In order to counteract the declining point value, individual budget caps (the so-called doctors' practice budgets) were introduced in 1997. Although these limited the reimbursable points per patient, the way in which they were set *ex ante* had to be abolished following a decision of the Federal Court of Justice in July 2003 which questioned the validity of the database used in the calculations. Whilst there has been no scientific evaluation on the impact of practice budget, it is fairly likely that they stabilized the point values (i.e. the monetary value of ambulatory physician services in general) as they did not decline any further after 1997 [20].

On 1 April 2005, a new EBM was passed after several years of preparation. One central intention of this new so-called EBM 2000plus was to further counteract a declining point value and also to alleviate the prisoner's dilemma situation of ambulatory physicians. The central elements of the new EBM 2000plus are as follows:

- New service complexes which combine several previous single fees for services; for example, the new EBM 2000plus includes in the physician patient contact complex several fees which were reimbursed separately in the old EBM. Cost-accounting methods have been used in order to calculate these service complexes.

- The service complexes have been calculated according to workload and time used by physicians. The service complexes are also adjusted for age.

- There is a strict separation between family doctors and specialists, where family doctors are not allowed to be reimbursed for specialist services, and *vice versa* ([21]: 48; [22]).

So far, only one preliminary evaluation has been made of the impact of the EBM 2000plus in two selected regions. The analysis covers the period of the four quarters before, and the four quarters after, the introduction of the EBM 2000plus, and revealed requirements for increases in total services by 8.5% and in services per case by 7.4% ([22]: 150).

5.2.1.3 Hospital Care

Whereas the reimbursement of physicians is a two-stage process within joint self-government (as noted above), there is no such procedure in the inpatient sector. One reason for this is that, for legal reasons, regional hospital federations lack the formal authority to distribute payments to individual hospitals. Consequently, the global budgets have not been enforced on a regional level but rather on the individual hospitals' level. Whether the current gradual replacement of these hospital budgets through a system based on diagnosis-related groups (DRGs) (for further information on this replacement process, see below) will lead to *de facto* budgets on a Federal State wide level – or to the abolition of budgets as anticipated by the hospitals – remains to be seen.

Since 1985 there have been three major reform steps in the way that hospitals are reimbursed (cf. for the following [4]: 140ff., in some parts verbatim). In chronological order these reforms have become increasingly comprehensive in their approach to changing the payment method of hospitals. Until 1985, hospitals were paid by *per diem* (per day) which were set for each hospital by the Ministries of the Federal States. The so-called 'full cost cover principle' was applied – that is, the hospitals were eligible for the full reimbursement of their costs.

In 1985, the method of payment was changed slightly with the introduction of prospective budgets that were negotiated between the hospital owners and the sickness funds. These budgets were reimbursed by *per diem* and procedure fees (the latter covering the expensive costs incurred in the operating room). *De facto*, however, hospitals were still reimbursed for their full costs, as any costs which were lower or higher than those negotiated were accounted for in the following year. Yet, this minor change in the payment method did not lead any to cost savings in hospitals [23, 24].

In 1992, a more far-reaching reform concerning the payment method of hospitals was introduced with the passing of the Health Care Structure Act. With this Act the full cost cover principle was legally abolished and budgets were calculated on an individual hospital level, initially for the period 1993 to 1995. Increases in the budget were limited to increases in the growth rate of the contributory income to the sickness funds. Although these budgets were intended to last only until 1995, some are still in use today. However, whereas budgets between 1993 and 1995 were fixed, those since then have been based on negotiated targets between hospitals and sickness funds, with a certain compensation for deviations from the target. The new payment method had two components. As of 1993, the hospital services were reimbursed by a two-tier system of *per diem* charges: (i) a hospital-specific basic *per diem* covering nonmedical costs; and (ii) a department-specific *per diem* covering medical costs, including nursing, pharmaceuticals and procedures [23]. The second component was made mandatory in 1996, when case fees (covering the costs for the entire hospital stay of a patient) and procedure fees (paid on top of slightly reduced *per diem* payments) were introduced in order to produce a more performance-related payment method for hospitals. However, case fees and procedure fees accounted for the reimbursement of less than one-quarter of all hospital services. Overall, there were about 70 case fees and 140 procedure fees, with the former being distributed unevenly between specialties. For example, there were no case fees for medical, pediatric and psychiatric patients, whereas more than 50% of cases in gynecology and obstetrics and approximately two-thirds of ophthalmologic cases were reimbursed via case fees [23].

Due to this new incentive structure, the following effects were anticipated (in particular in areas were case and procedure fees were applied): increased efficiency of service provision because of more cost-consciousness; cream skimming; (too) early discharges of patients who are paid via case fees; a decrease in the average length of stay in general; and increased service specialization of hospitals, as they will concentrate on case fees which they can provide profitably. Although scientific evidence is rather scant on whether these expected effects have materialized, there is some evidence that the hospitals introduced various measures to increase efficiency [25], and that the overall efficiency of the hospital sector improved during the period 1991–1996 [26].

The original intention of the government gradually to extend the scope of services reimbursed via case fees to 100% did not materialize, however, and in the year 2000 a new approach was chosen. The government committed joint self-government (in this case: the German Hospital Federation and associations of the statutory sickness funds and private health insurers) to introduce a universal (i.e. for *all* hospital services with the major exceptions of psychiatry and psychosomatic medicine) performance-related case fees system based on DRGs. The self-governing bodies opted for the Australian Refined DRG system and adapted it into the German-DRG (G-DRG) version, which is modified annually but will only gradually become fully effective

Table 5.5 German DRG system: key indicators: 2003–2008.

	2003	2004	2005	2006	2007	2008
DRGs total	664	824	878	954	1082	1137
DRGs for main department (valued)	642	806	845	912	1035	1089
DRGs for main department (unvalued)	22	18	33	40	42	43

Source: Refs [8]: 940; [9]: 1074.

as a price system. In the so-called *convergence phase* (2005–2010), hospital budgets will be incrementally transformed towards a uniform price level which is valid on a Federal State-wide level. The possibility of losses for hospitals due to adjustments towards the DRG system was legally restricted. In 2005, a hospital could lose 1% of its budget at most because of the convergence phase, but this is increased to 3% in 2009. The possibilities for losses are unbounded, however, beginning with the year 2010 ([21], cf. this reference also for more information on the G-DRG system and the convergence phase). Until then, the final legal frame of reference for the price system must be set which will condition, to a large degree, its incentives effects.

The data in Table 5.5 show that the number of reimbursable G-DRGs rose quite considerably between 2003 and 2008. However, more recent versions of the DRG system cannot account for the total variation in the costs of hospitals. The classification of the G-DRG-version 2007 could explain approximately 71% of the variance (measured by R^2) of costs (based on data for 2005) on the basis of all cases, and approximately 81% on the basis of so-called 'inliers' ([27]: 14). Nonetheless, a classification system which explains a maximal amount of cost variation is necessary, in order to achieve an allocative efficient system. In order to accomplish this goal, hospitals will most likely have to improve their accounting systems. An analysis of the G-DRG-version 2006 revealed a very high correlation between cost weight and mean length of stay. This is probably the case because hospitals have – so far – no more precise allocation criteria ([28]: 276). The hospital-specific base rates for 2003–2006[7] shown in Table 5.6 indicate that there is no obvious trend towards less dispersion (which one would expect).

5.2.1.4 Financial and Other Performance Indicators of SHI Ambulatory Physician Practices and Hospitals

In this section we turn to indicators which describe performance on the level of individual physician practices and hospitals (though mainly based on aggregate data), the aim being to infer some consequences of the strategies described above.

---

7) These hospital-specific base rates are lower than average adjusted costs to sickness funds presented in Table 5.8. This is because psychiatric hospitals are included in the latter, where expenditure is much higher per case.

Table 5.6 Hospital-specific base rates: average and dispersion (2003–2006), rounded to nearest Euro.

Year	Average base rate (€)	No. of hospitals	SD	Min	Max
2003	2546	999	371	1310	4158
2004	2624	1673	407	961	6200
2005	2717	1663	347	1347	5582
2006	2739	1566	381	1588	10 116

Source: http://www.aok-gesundheitspatner.de/bundesverband/krankenhaus/budgetverhandlung/basisfallwerte/ (accessed 29 November 2007) and own calculations.

The data in Table 5.7 show that there was a quite substantial increase in the number of SHI-affiliated physicians in the ambulatory sector between 1996 and 2005. The increases of total remuneration for all ambulatory physicians only made up for this increase in the number of physicians. This development contributed to place pressure on the income for individual physician practices. As shown in Table 5.7, the average remuneration per SHI physician practice declined (in nominal terms!) between 1996 and 2005. Expenditure per case increased by about 13% between 1996 and 2004 (due mainly to the new EBM 2000plus in 2005, figures for this year are not comparable with those of previous years).

Table 5.8 presents some indicators for the hospital sector for the period 1991 to 2005. Costs to sickness funds for hospitals per case rose, notably between 1992–1995

Table 5.7 Indicators for the ambulatory care by SHI-affiliated physicians; changes in the number of physicians, services provided and remuneration 1996–2005 (in current prices).

Year	No. of SHI-affiliated physicians[a]	Remuneration for all physicians (billion €)	Remuneration per SHI physician (€)	Expenditure per case[b] (€)	Expenditure per insured member (€)
1996	107 071	20.1	188 046	39.6	396.3
1997	108 734	20.4	187 868	39.0	401.9
1998	110 339	20.6	186 743	38.7	406.7
1999	122 604	21.7	176 744	39.3	425.7
2000	126 487	22.5	177 614	40.3	440.3
2001	128 333	23.2	181 003	41.1	455.5
2002	131 251	23.8	181 430	41.6	467.2
2003	129 950	24.2	186 066	41.5	476.4
2004	130 278	24.1	184 996	44.6	476.1
2005[c]	133 239	24.8	186 153	51.6	492.0
Change in % 1996–2005	24	23	−1	30	24

Source: Ref. [6]: 43 and own calculations.
[a]From 1999 including psychological psychotherapists.
[b]A case is defined as one or more patient contacts with one and the same physician per quarter.
[c]Modified measuring and billing due to introduction of new the EBM 2000 plus.

Table 5.8 Performance indicators of hospitals (1991–2005).

Year	Average adjusted cost to sickness funds (€)						Length of stay and occupancy rate			
	Per bed	% change to previous year	Per day	% hange to previous year	Per case	% change to previous year	ALOS (days)	% change to previous year	Occupancy rate (%)	% change to previous year
1991	56 224		183		2567		14.0		84.1	
1992	63 782	13.4	208	13.3	2756	7.4	13.2	−5.7	83.9	−0.2
1993	68 826	7.9	227	9.3	2848	3.3	12.5	−5.4	83.1	−1.0
1994	73 195	6.3	243	7.2	2920	2.5	11.9	−4.8	82.5	−0.7
1995	78 549	7.3	262	7.7	3003	2.8	11.4	−4.2	82.1	−0.5
1996	81 448	3.7	276	5.3	2992	−0.4	10.8	−5.3	80.6	−1.8
1997	83 878	3.0	283	2.7	2963	−1.0	10.4	−3.7	81.1	0.6
1998	86 821	3.5	289	2.0	2946	−0.6	10.1	−2.9	82.3	1.5
1999	89 514	3.1	298	3.2	2960	0.5	9.9	−2.0	82.2	−0.1
2000	92 207	3.0	308	3.1	2989	1.0	9.7	−2.0	81.9	−0.4
2001	95 788	3.9	324	5.3	3056	2.2	9.4	−3.1	81.1	−1.0
2002	99 976	4.4	342	5.7	3139	2.7	9.2	−2.1	80.1	−1.2
2003	102 721	2.7	363	6.0	3218	2.5	8.9	−3.3	77.6	−3.1
2004	105 633	2.8	382	5.5	3341	3.8	8.7	−2.2	75.5	−2.7
2005	108 304	2.5	392	2.6	3362	0.6	8.6	−1.1	75.6	0.1

Sources: Table 1.1 in Ref. [29]; Table 1.1 in Ref. [7] and own calculations.

and 2001–2004, but then *decreased* during the period 1996–1998. The start of the reduced expenditure per case coincided not only with the introduction of prospective case fees (from 1 January 1996) but also with the introduction of institutional benefits from the long-term care insurance (from 1 July 1996). There is some evidence that patients were transferred more frequently and earlier to rehabilitation clinics [25], and that 'costly' patients were transferred more frequently to university hospitals that have virtually no possibility to transfer such patients, as they are the 'providers of last resort' [25, 30]. The average length of stay decreased disproportionately in departments where case fees were applied above average [23, 25]. As the data in Table 5.8 show, the overall average length of stay decreased during the whole period covered. Decreases were particularly pronounced in those years when fixed budgets for hospitals (1993) and prospective case fees (1996) were introduced. On the other hand, the decrease from 2003 to 2005 in the average length of stay (i.e. since the introduction of the G-DRGs) was not particularly strong.

5.2.2
Cost Shifts to Private Households

The information provided in Table 5.3 shows that out-of-pocket payments as a share of total health expenditure rose rather moderately during the 1990s. A noticeable increase subsequently occurred, in particular between 2003 and 2004. These increases are not some type of volume effect, but most likely are mainly rooted in increases of out-of-pocket payments and benefit exclusions on the individual level. These increases were particularly pronounced in the years 1993, 1996–1998 and 2004, exactly when the benefit exclusions took place [31].

The health care reform which shifted costs to private households most heavily was the SHI Modernization Act (SHIMA), which was enforced from 1 January 2004. Among others changes, it introduced a user charge of €10 for the first contact with ambulatory physicians and dentists, per quarter. Moreover, further ambulatory physician contacts are also charged with €10, except when the patient has a referral and the €10 charge is waived. So, the new measure clearly involves an incentive for the patient to obtain a referral after the first physician contact per quarter (for more on the increases of cost-sharing measures of this act, see [1]: 198ff.; [31]: 28f.).

Very little research has been conducted on the consequences of these increases in cost-sharing on issues such as allocative efficiency or the utilization of health care services. One study evaluated the impact of the user charge for ambulatory physicians by comparing the latters' consultations in 2003 and 2004. Although compared to 2003 the number of consultations fell significantly in 2004, there was no evidence that this reduction was caused by a decrease in consultations by people with disabilities or in ill health; rather, the results indicate that medically unnecessary consultations were avoided [32]. These results were confirmed by two representative surveys among SHI insurees in 2004 and 2005, which showed that referrals by GPs increased following the introduction of the user charge for ambulatory physicians, and that there were no social distortions. This would mean that the intended impact – an increased steering function of GPs – did in fact materialize through the reform [33]. However, the results of a different survey which contained information on physician contacts up to autumn 2005 indicated that the utilization of ambulatory physicians in 2005 was higher than in 2003 and 2004 ([34]: 29–31). Furthermore, a different study showed that the measures of the SHIMA led to an underutilization of pharmaceuticals among the particular vulnerable group of homeless people ([35]: 721). Previous research findings suggests, then, that health care utilization behavior is both dynamic and complex, and deserves further scrutiny. For example, price increases in Germany were particularly pronounced during recent years for people with low incomes, and this in turn could have affected their physician and pharmaceutical utilization behavior.

Two other aspects are deserving of mention at this point. First, even after increases of cost sharing of the SHIMA, the proportion of expenditure of private households ranks average by international comparison [36]. Second, the SHIMA is the first Act to substantially increase the private share of health care financing more than any other single health care reform act. This might indicate that it has become politically more difficult to shift the burden of cost containment mainly onto providers via budgets (which may also have an indirect impact upon private households, but this is even more difficult to measure than direct increases of copayments).

5.2.3
Promoting Competition between Sickness Funds and Providers

Traditionally, the SHI has been an occupationally stratified health insurance; this means that, for most employees, there was no choice of statutory health insurance. Prior to 1996, affiliation to a statutory sickness fund rested upon an assignment system, which was based on occupation, respectively employment, in a certain company. In particular, the following rules applied:

- The AOKs were the so-called 'basic sickness funds', and were responsible for all compulsory insured unless they were not obliged to be insured elsewhere.
- If a company had its own BKK, then this took over the responsibility of the AOK. If the company belonged to a guild, then the employees of this company were assigned to the respective guild fund.
- The substitutive funds did not have a mandatory insurance clientele, but were able to open themselves for certain occupational groups.
- In addition, there were three special insurance systems for farmers, miners and sailors ([37]: 14f.).

According to the principles named above, the majority of insured people had no choice over their sickness fund and were assigned to the appropriate fund based on geographical and/or job characteristics. This mandatory distribution of fund members led to greatly varying contribution rates because of different incomes and risk profiles. Only voluntary white collar members – and since 1989 also voluntary blue collar members – had the right to choose among several funds and to end their membership with two months' notice. Other white collar workers (and certain blue collar workers) were allowed to choose when becoming a member or changing jobs.

As this group grew substantially over the decades, around 50% of the population had at least a partial choice during the early 1990s ([1]: 60).

In 1996 this situation changed completely, however, when the vast majority of SHI members (ca. 97%) were allowed to freely choose their sickness funds. Forced assignment was retained only for the miners', farmers' and sailors' funds. Incentives for risk selection were counteracted by the introduction of a risk adjustment scheme in 1994 (without pensioners, but since 1995 including pensioners). Until the year 2002 only the following risk-adjusters were in force: age, gender, entitlement to sickness allowance, and receipt of disability insurance (and also equalization of contributory income). Two important measures were passed in order to improve the risk-adjustment scheme. First, a high-cost pool has been introduced which compensates sickness funds for insurees whose health care costs exceed a threshold of € 20 450 annually: 60% of these costs are compensated, while the respective sickness fund has to cope with the remaining 40% on its own [38].

Second, Disease Management Programmes (DMPs) were introduced. These are a managed-care instrument and can be offered by statutory sickness funds if the underlying indication was accredited by the Federal Insurance Office (for more information on the introduction of DMPs, see [39]). In June 2007 the following DMPs were accredited by the Federal Insurance Office, and can therefore be offered by the sickness funds: diabetes mellitus types 1 and 2, coronary heart disease, asthma/chromic obstructive pulmonary disease (COPD) and breast cancer. DMPs are linked to the risk-adjustment scheme since DMP enrolees form separate categories in the scheme which, in almost all cases, leads to higher transfer payments for the respective sickness fund.

The free choice of insurees and the concomitant introduction of the risk adjustment scheme led to, or even reinforced, several developments. First, the number of sickness funds dropped remarkably. Second, the contribution rates of sickness funds converged. Third, due to the increased mobility of insurees between sickness funds, the market shares of the groups of sickness funds changed considerably.

1. In 1993 – some three years before the introduction of complete free choice for insurees – there were 1221 sickness funds, but this number fell to 242 in 2007 (Table 5.9).[8] This decline is not due to the closure of sickness funds, but to mergers, the first wave of which took place between 1994/1995 and affected the general regional funds. In 1995 the guild funds followed. The number of company-based funds steadily declined, with a pronounced drop between 1995 and 1997, directly before and during the introduction of full competition. The latest health care reform which came into force on 1 April 2007 allowed mergers between sickness funds belonging to different groups, which should facilitate even more mergers in the future. Several reasons have been proposed for the mergers, with perhaps the most important from the early to mid-1990s being the pooling of risk structures of sickness funds. One reason with increasing

---

8) Historically there have been a large number of sickness funds; before 1911 there were about 22 000 ([62]: 6).

Table 5.9 Number of statutory sickness funds between 1993–2007 (1 January).

Fund	1993	1995	1997	1999	2001	2003	2005	2007
General regional funds	269	92	18	17	17	17	17	16
Company-based funds	744	690	457	361	318	260	210	189
Substitute funds	15	15	14	13	12	12	10	10
Guild funds	169	140	43	42	28	23	19	16
Farmers' funds	22	21	20	20	19	10	9	9
Sailors' fund	1	1	1	1	1	1	1	1
Miners' fund	1	1	1	1	1	1	1	1
Total	1221	960	554	455	396	324	267	242

Source: Table 9.4 in Ref. [40]; also Refs [41]: p. 37, [5]: p. 37.

relevance is a gain in the bargaining power of sickness funds in negotiations with providers. Related to this is the build-up of competencies in care management, as sickness funds (to an increasing degree in Germany) are not only payers but also players of care – that is, they are increasingly involved in the planning and management of health care (see also below).

2. Before the introduction of the free choice of sickness fund and risk-adjustment scheme, the range of contribution rates between some single sickness funds was 8.8 percentage points (minimum 8.0% and maximum 16.8%) [42]. Both, the range between minimum and maximum contribution rate of individual sickness fund and variance between the contribution rates of sickness fund groups declined since 1996 (Figure 5.2). On 1 July 2006, the range between the sickness fund with the highest and lowest contribution rate was 3.3 percentage points (minimum 11.3% and maximum 14.6%) [43].

3. The possibility to choose a sickness fund freely was exercised to a considerable degree. According to the research findings of Andersen and Grabka ([44]: 180f.), three periods of change of sickness funds can be distinguished. In the first period (1996–1998) there was a variety of motives for change, the most important being the amount of contribution rate, the image of the sickness fund, and perceived differences in service or benefits. The second period (about 1998–2002/2003) was almost exclusively dominated by the amount of the contribution rate as the reason for change. During this phase, there was a clear increase in the number of people who changed sickness funds several times in order to find the fund with the lowest contribution rate at the time. The number of people who changed funds for the first time stagnated, and subsequently decreased, during this period. There are indicators which suggest that a new period began around 2002/2003, when the numbers of people changing funds rose again. Those people who change sickness funds several times appear to accept higher contribution rates of new funds, probably because they expect a better management of their needs by the new funds. Moreover, and compatible with this, the perceived subjective health of people who changed sickness fund several times declined slightly ([44]: 181).

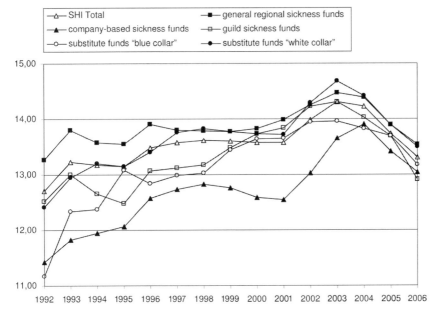

Figure 5.2 Contribution rates of groups of sickness funds.
Note: sailors' and miners' sickness funds are not included
([5]: 1 and own calculations).

An analysis of the determinants of switching behavior of insurees, using data from the German Socio-Economic Panel, did not produce any evidence for risk selection by *groups* of sickness funds, for example general regional funds, company-based funds, and so on [45]. This result does not foreclose that risk selection is in fact practiced by some *individual* funds, and there is indeed some anecdotal evidence available for this [46].

Company-based funds almost doubled their membership between 1995 and 2006, whereas the general regional funds lost 8 percentage point (Figure 5.3). The success of the company-based funds originates in incomplete risk adjustment, together with a negative correlation between health status and switching costs ([45]: 1267ff.). If the analysis of Andersen and Grabka ([44]: 180f.) is correct, then this latter aspect will be of declining importance in the future.

As noted above, during the starting period of virtually complete competition between sickness funds, the amount of contribution rate played an important role in the decision to choose a particular fund. Since about the late 1990s (however, with a noticeable impact on the behavior of sickness funds and providers only with the passing of the SHIMA in 2004) the Government has gradually enhanced the possibilities for individual sickness funds to selectively contract with providers or groups of providers in order to stimulate competition, not only on the amount of the contribution rate but also on specific arrangements of care. The two most important measures of this type are the above-mentioned DMPs and the possibility for sickness funds to conclude integrated care contracts.

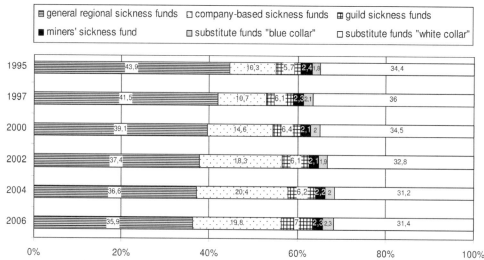

Figure 5.3 Proportion of market shares of groups of statutory
sickness funds (1995–2006). *Note:* until 2004 annual averages;
2006: 1 July; sailors' and farmers' sickness funds are not included.
(Table 9.7 in Ref. [47] for 2006: Bundesministerium für
Gesundheit 2006).

DMPs can be seen as special integrated care contracts, as they are linked to the risk-adjustment scheme. Therefore, sickness funds have strong financial incentives to subscribe their insurees (who fulfill the medical criteria) to their DMPs. In contrast to integrated care contracts, DMPs are heavily regulated and must be accredited by the Federal Insurance Agency. Preliminary evidence points to the fact that they have an improved quality of care [48], although the results of the very few relevant and properly conducted trials (i.e. with a control group) are still awaited. However, when the criticism was made that people who might benefit disproportionately from DMPs were under-represented in them, the question was raised as to whether the improvements in quality are cost-effective ([35]: 236–241).

The possibility to conclude integrated care contracts has existed since the instigation of the Health Care Reform Act 2000 ([49]: 861). Initially, however, the bureaucratic hurdles were too high, and so the associations of SHI physicians exerted veto power, as they stood to lose influence if integrated care contracts gained importance. On coming into force in 2004, the SHIMA eased the process and also provided a financial incentive for the conclusion of such contracts (for more details, see [50]). During the first quarter of 2007, a total of 3498 such contracts existed (with 4.07 million subscribed insurees in these integrated care arrangements). Most of these referred to specific indications, with about 25% concerning hip and knee joint endoprostheses ([35]: 221–224). So, the SHIMA constituted a breakthrough with regards to the conclusion of such contracts. To date, it is still unclear as to how sustainable these contracts will be when the legally provided financial incentives are abolished at the end of 2008, and to what degree they will in fact promote a better integration of care.

5.2.4
Strengthening the Competencies of Joint Self-Government

As described briefly in Section 5.1, the Federal Joint Committee is the central body in relation to joint self-government, despite being founded only in 2004. Its main predecessor was the Federal Committee of Physicians and Sickness Funds (Bundesausschuss der Ärzte und Krankenkassen), but this body (founded in 1955) was only responsible for the ambulatory sector. In the year 2000 this institution was complemented by a committee for the inpatient sector which performed similar functions for hospitals, in addition to a coordination committee with the function to link the two sectors.

The SHIMA unified these institutions under a new umbrella agency, the Federal Joint Committee (Gemeinsamer Bundesausschuss; GBA), which is also the legal successor of these institutions (the following is mainly based on Ref. [51]: 49–51). The SHIMA also extended the competencies of joint self-government. The extension of competencies to the Federal Joint Committee continues to this day; for example, it was made responsible in 2007 to work out the details of specialized palliative ambulatory care, and which highly specialized ambulatory treatments may be provided in inpatient settings.

The Federal Joint Committee decides on many aspects of the benefit catalogue of the SHI, and also assesses new methods of medical examination and treatment. In addition, it defines evaluative criteria for quality assurance measures for the different sectors of the health care system (http://www.g-ba.de/cms/front_content.php?idcatart=207&lang=1&client=1).

The Federal Joint Committee has 30 members (Figure 5.4) and, for the first time, representatives of patients are included within the structures of self-government. However, they do not have the same rights as the representatives of the sickness funds and the providers. While they have the right to make applications and take part in the proceedings of the Federal Joint Committee, they have no voting rights in the decision making [52]. Most decisions of the Federal Joint Committee are made in the subcommittees, where the Federal Joint Committee meets in varying compositions (though always with the three parts – sickness funds, providers and patients).

To assist the Federal Joint Committee in its tasks, the Institute for Quality and Efficiency in Health Care (IQWiG) was founded in the autumn of 2004 (http://www.iqwig.de/). The institute is firmly integrated in the system of self-government, and is financed by the associations of the sickness funds, physicians and hospitals. The institute has, among others, the following tasks:

- Research, descriptions and evaluations of the up-to-date standard of medical knowledge for diagnostic and therapeutic technologies for selected diseases.
- Compilation of scientific reports and statements on questions relating to the quality and efficiency of the services are provided by the SHI.

As indicated above, in relation to pharmaceuticals, the economic evaluation of health technologies will become more important in the future. The latest health care

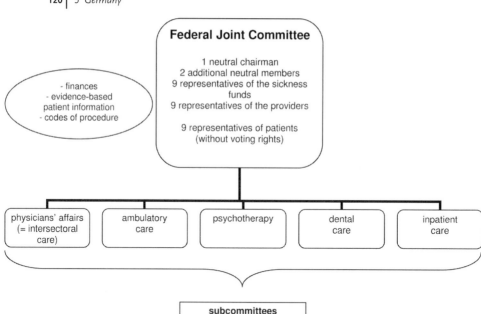

Figure 5.4 The structure of the Federal Joint Committee. (Based on Bronner [52] and http://www.g-ba.de/downloads/17-98-2436/ 2007-09-05-G-BA_im_Ueberblick.pdf (accessed: 06/9/2007).

reform which was passed in February 2007 made it possible for the IQWiG to evaluate the relationship of costs and benefits of pharmaceuticals, and also of other health technologies. The IQWiG has been commissioned to develop an analytical framework for the economic evaluation of drugs and other interventions. In January 2008, the IQWiG – together with an International Expert Panel – produced a first version of this analytical framework, the Institute for Quality and Efficiency in Health Care (2008) [53].

Two other organizations must be mentioned at this point if the increasing role of joint self-government is to be discussed, both of which are administered by the joint self-government:

- The Institute for Hospital Reimbursement (Institut für das Entgeltsystem im Krankenhaus; InEK) founded in 2001, which is there to provide the organizational structure to maintain and further develop the G-DRGs as a reimbursement system ([28]: 272).

- The Federal Office for Quality Assurance (Bundesgeschäftsstelle Qualitätssicherung; BQS), which was founded in 2000 and began working in 2001. The BQS supports the statutory system in the development and implementation of measures for external quality assurance in hospitals, as stipulated in the Social Code Book.

Both organizations also cooperate with the Federal Joint Committee.

5.2.5
Summarizing and Concluding Remarks on Cost Containment and Efficiency Seeking Strategies

Four main strategies have been identified to contain expenditure and promote efficiency in the preceding sections: (i) the imposition of global budgets and spending caps; (ii) cost shifts to the private sector; (iii) increased competition between statutory sickness funds; and (iv) increasing the role of joint self-government.

Since the early 1990s, the core measures to promote cost containment have been the setting of budgets for providers, in particular for hospitals and ambulatory practices. Cost shifts to private households as a strong means to relieve SHI expenditure have been used in 2004 (and compared to the shifts passed in 2004 rather moderately before that). It is highly plausible that, in particular, these budget settings and budget shiftings were responsible for the fact that SHI expenditure remained virtually stable as a share of GDP. Moreover, as was shown, the expenditures for the two sectors which were primarily affected by budgets (hospitals and physician's practices) remained stable as a share of GDP. This can be considered a success of cost-containment strategies. The consequences of these strategies, however, have been an increasing dissatisfaction of providers and political pressure against budget setting. It appears that an attenuation of budgets in the ambulatory and inpatient sectors have already been initiated (see Section 5.3 for more information on this).

Cost containment on its own is rather an absurd aim. What is aspired to is a good balance of affordability, acceptability, comprehensiveness and effectiveness. The increasing emphasis on cost containment is rooted in the notion that health care systems are not sufficiently cost-effective ([54]: 482). In 1991, Evans and colleagues wrote, with reference to Canada and the USA:

> "Does cost containment have a cost in terms of health? Are Canadians healthier because their access to care is unimpeded by financial barriers, or are Americans healthier because they spend more on care? And which Canadians/Americans? The American without insurance? The Canadian without a coronary artery bypass graft? Most of these difficult questions (whom to count, how to weight and aggregate people), are moot because so little is known, in any country, about either the health status of the population or the effectiveness of health care interventions, let alone the connection between the two. *One might have hoped that successful cost containment would have been associated with a rapid expansion of such knowledge, because when less is spent it is more important to spend effectively. One would be disappointed*" ([54]: 497, our emphasis).

More than 15 years later, and with reference to Germany, one can state that increasing use is made of health technology assessment (HTA) and economic evaluation tools to back up coverage decisions for single technologies. However, there is no evidence or scientific effort which tries to infer the consequences of cost containment on health or health care in general (which might be methodologically impossible to do). One is left, therefore, to look at the general broad macro indicators.

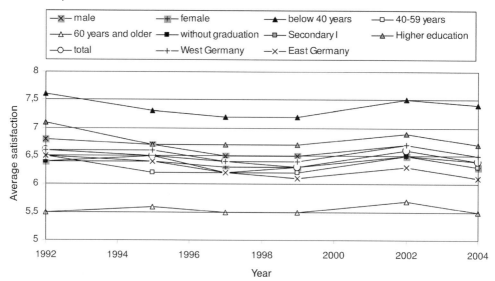

Figure 5.5 Satisfaction with health status (average satisfaction).
Note: *Average satisfaction* is the average of the answers (answers
were possible on a scale between 0 and 10, where 10 indicates
complete satisfaction with health and 0 is complete
dissatisfaction) [55–57].

First, in spite of cost-containment policies, life expectancy at birth increased and
still continues to increase (for more information, see Ref. [4]: S145). Second, as
shown in Figure 5.5, self-assessed health – which is an acceptable proxy for objective
health – did not increase but was inconsistent between 1992 and 2004 (these data are
derived from a panel sample survey which is representative of the German population in total). However, one cannot relate these data to the health care system or health
care policies. As far as confidence in the German health care system is concerned, it
must be stated that confidence declined between 1996 and 2002 [58].

While budget setting and shifting directly impacted upon SHI expenditure, many
structural reforms have been introduced since the 1990s which gradually altered the
structure and regulation of the SHI system. Among the most prominent of the
reforms is the introduction of entirely free competition between sickness funds,
which has resulted in particular in four developments: (i) a decline in the number of
sickness funds; (ii) an increased mobility of insurees; (iii) a convergence of contribution rates; and (iv) changes in market shares between groups of sickness funds.
The Government embellished these competitive arrangements between sickness
funds by successively providing them with more possibilities to compete and to gain
profiles distinct from other sickness funds.

An important component of this competitive strategy is the possibility for sickness
funds to conclude integrated care contracts with health care providers. As shown
above, although approximately 3500 such contracts were concluded it affected only a

small percentage of insurees. It became evident that the conclusion of such contracts is connected with many difficulties. Besides certain legal hurdles (most of which have been removed in the course of time), veto power was exerted by the Federal and Regional Associations of SHI Physicians, as they lose influence with the increasing importance of such contracts. In addition, these contracts need a changed role perception of individual sickness funds as managers of care and increased managerial and administrative competencies on both sides – sickness funds and providers. In the future, individual sickness funds will have to negotiate much more with individual providers (or amalgamations of providers) on the terms and conditions of care than is presently the case. As shown in Table 5.1, many ambulatory physicians work in either solo or small practices, and normally do not have the capabilities to engage in rather complex contract negotiations with sickness funds.

At the same time, competencies of the joint self-government were extended which have led to an ever more regulated SHI system, in particular in relation to the benefit catalogue and aspects of quality assurance.

Almost all of the described reform measures concern the organizational and expenditure side of the SHI. Figure 5.6 shows that normal increases of SHI expenditure exceed increases of SHI revenue, but regular cost-containment acts (normally once every legislative period) inverse this relationship for a short time. However, Figure 5.6 also illustrates a more fundamental problem (this and the following is based on Ref. [59]: 81, where this problem is discussed in much more detail). Without these different cost-containment acts the share of SHI expenditure on GDP would have clearly risen. The revenue basis of the SHI is too weak in order to guarantee stability of contribution rates (cf. Figure 5.2, which shows increasing

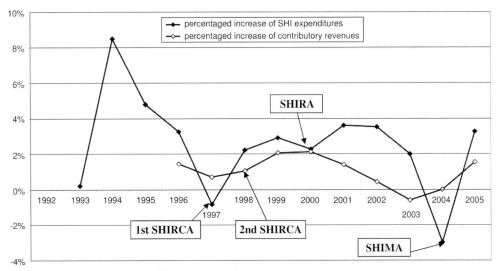

Figure 5.6 Percentage increases of SHI increases and contributory income 1992–2005. ([5], 2 and own calculations).

contribution rates). If this trend were to continue in the future then the SHI would face the following dilemma: If the expenditure of SHI is linked to increases in GDP, then contribution rates will inevitably also increase. The principle of contribution rate stability makes it necessary that increases in SHI expenditure are lower than increases in GDP, although the latter places severe constraints on the financial situation of the SHI.

The latest health care reform now tries to tackle this problem on the financing side with the introduction of the new health fund. This is a core element of the current reform situation which will be described in Section 5.3.

5.3
Report on the Current Situation

This section will deal primarily with the situation after the SHI-Competition Strengthening Act (SHI-CSA), which was passed on 16 February 2007. The SHI-CSA initiated the creation of a completely new institution – the health fund – which will come into operation on 1 January 2009. As shown in Figure 5.7, starting with the year 2009, sickness funds will lose their competence to determine individually the amount of the contribution rate. Instead, a uniform contribution rate for all sickness funds will be determined by the Federal Ministry of Health and issued as an ordinance. In the future, sickness funds will obtain their revenue from the health fund according to a flat amount per insuree which is adjusted for age, gender and sickness.[9] If the individual sickness fund cannot operate (on a financial basis) with the revenue from the health fund, it will have to levy an additional charge on its insurees. But, this additional charge must not exceed 1% of the contributory income of insurees. As the scheme in Figure 5.7 shows, the Health Fund provides means to (and intends to) inject tax money into the SHI system which can contribute to alleviate the disaccord between SHI expenditure and total contributory income.

Although, in future, sickness funds will not be able to decide upon the amount of their general contribution rate, competition between funds will be intensified. There will be a strong incentive for sickness funds not to levy an additional charge on their insurees [61].

The SHI-CSA further promotes competition between sickness funds by forcing them to offer different tariffs (e.g. cost sharing) and GP gate-keeping models for their insurees. The reform act also made an important step in relation to the reform of ambulatory reimbursement. As shown above, although reimbursement in the ambulatory sector was marked by floating point values, this is an unstable financial situation for outpatient physicians. The reform strives for a new fee schedule which contains fixed euro values.

9) In the year 2009 a new morbidity-orientated risk-adjustment scheme is due to start that will adjust for about 50–80 complex and cost-intensive diseases.

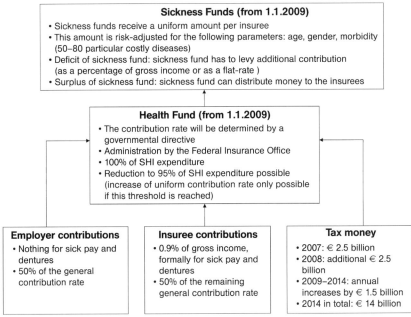

Figure 5.7 The health fund as of the year 2009. ([60], slightly modified, translated and amended by the authors and Miriam Blümel).

So, it might be that the SHI-CSA will be considered in future as an important step which shifted the focus away from restrictive budgets towards more flexible forms of cost containment. In future, the Government will have more discretion in deciding on the sources of health care finance, notably in relation to the mix of tax money, contributions and expenditure of private households.

References

1 Busse, R. and Riesberg, A. (2004) *Health Care Systems in Transition: Germany*, WHO Regional Office for Europe on behalf of the European Observatory on Health Systems and Policies, Copenhagen.

2 Bundesministerium für Gesundheit und Soziale Sicherheit (Hrsg.) (2004) *Übersicht über das Sozialrecht*, BW Bildung und Wissen, Nürnberg.

3 Sachverständigenrat zur Begutachtung der gesamtwirtschaftlichen Entwicklung (2004) *Erfolge im Ausland – Herausforderungen im Inland. Jahresgutachten 2004/05*, Wiesbaden.

4 Wörz, M. and Busse, R. (2005) Analysing the impact of health-care system change in the EU member states – Germany. *Health Economics*, **14** (Suppl. 1), 133–149.

5 Bundesministerium für Gesundheit (2007) Gesetzliche Krankenversicherung. Mitglieder, mitversicherte Angehörige, Beitragssätze und Krankenstand. Monatswerte Januar bis Juli 2007. (Ergebnisse der GKV-Statistik KM 1).

6 Kassenärztliche Bundesvereinigung (2007) *Grunddaten zur Vertragsärztlichen Versorgung in der Bundesrepublik Deutschland 2006. Zahlen, Fakten,*

Informationen, Kassenärztliche Bundesvereinigung, Dezernat 4 – Versorgungsstruktur und veranlasste Leistungen, Berlin.

7 Statistisches Bundesamt (2006) Gesundheitswesen, Fachserie 12: Reihe 6.1.1: Grunddaten der Krankenhäuser – 2005 Wiesbaden.

8 Schlottmann, N., Fahlenbach, C., Brändle, G. and Wittrich, A. (2006) G-DRG-System 2007. Abbildungsgenauigkeit deutlich erhöht. *das Krankenhaus*, **98** (11), 939–951.

9 Schlottmann, N., Köhler, N., Fahlenbach, C. and Brändle, G. (2007) G-DRG-System 2008. *das Krankenhaus*, **98** (11), 1070–1082.

10 Hacker, J.S. (2004) Review article: dismantling the health care state? Political institutions, public policies and the comparative politics of health reform. *British Journal of Political Science*, **34**, 693–724.

11 Mossialos, E. and Le Grand, J. (1999) Cost containment in the EU: an overview, in *Health Care and Cost Containment in the European Union* (eds E. Mossialos and J. Le Grand), Ashgate, Aldershot, Brookfield USA; Singapore; Sydney, pp. 1–154.

12 Maynard, A. (1996) Efficiency of spending under fixed budgets, in *Fixing Health Budgets: Experience from Europe and North America* (eds F.W. Schwartz, H. Glennerster and R.B. Saltman), John Wiley & Sons, Ltd, Chichester, pp. 49–62.

13 Schreyögg, J. and Busse, R. (2005) Physician drug budgets in Germany and effects on prescription behaviour. *Journal of Pharmaceutical Finance, Economics & Policy*, **14**, 77–95.

14 Busse, R. and Howorth, C. (1996) Fixed budgets in the pharmaceutical sector, in Germany: effects on costs and quality, in *Fixing Health Budgets: Experience from Europe and North America* (eds F.W. Schwartz, H. Glennerster and R.B. Saltman), John Wiley & Sons, Ltd, Chichester, pp. 109–127.

15 Busse, R., Schreyögg, J. and Henke, K.-D. (2005) Regulation of pharmaceutical markets in Germany: improving efficiency and controlling expenditures? *International Journal of Health Planning and Management*, **20**, 329–349.

16 Stargardt, T., Schreyögg, J. and Busse, R. (2005) Arzneimittelfestbeträge: Gruppenbildung, Preisberechnung mittels Regressionsverfahren und Wirkungen, *Das Gesundheitswesen*, **67**, 469–478.

17 Petkantchin, V. (2006) *Economic Effects of Germany's Reference Pricing Policy for Drugs, Research paper*, Institut économique Molinari, Brussels.

18 Henke, K.D., Murray, M.A. and Ade, C. (1994) Global budgeting in Germany: lessons for the United States. *Health Affairs*, **13** (4), 7–21.

19 Busse, R. and Howorth, C. (1999) Cost containment in Germany: twenty years experience, in *Health Care and Cost Containment in the European Union* (eds E. Mossialos and J. Le Grand), Ashgate, Aldershot, Brookfield, USA; Singapore; Sydney, pp. 303–340.

20 Krauth, C., Schwartz, F.W., Perleth, M., Buser, K., Busse, R. and Schulenburg, J.-M.G.v. (1997) *Zur Weiterentwicklung des Vergütungssystems in der ambulanten ärztlichen Versorgung*, Forschungsstelle für Gesundheitsökonomie und Gesundheitssystemforschung, Hannover.

21 Schreyögg, J., Tiemann, O. and Busse, R. (2005) *Health Basket Work Package 6: Approaches for Costing and Pricing in Practice*, Department of Health Care Management, Berlin.

22 Zentralinstitut für die kassenärztliche Versorgung & Wissenschaftliches Institut der Ortskrankenkassen (2006) Wissenschaftliche Begleitung zur Einführung des EBM 2000plus. Abschlussbericht (Version vom 12. 2006) im Auftrage der Kassenärztlichen Bundesvereinigung und der Spitzenverbände der Krankenkassen Berlin; Bonn.

23 Busse, R. and Schwartz, F.W. (1997) Financing reforms in the German hospital sector – from full cost cover principle to prospective case fees. *Medical Care*, **35** (10), 40–49.

24 Simon, M. (2000) *Krankenhauspolitik in der Bundesrepublik Deutschland. Historische Entwicklung und Probleme der politischen Steuerung stationärer Krankenversorgung,* Westdeutscher Verlag, Opladen.

25 Asmuth, M., Blum, K., Fack-Asmuth, W.G., Gumbrich, G., Müller, U. and Offermanns, M. (1999) *Begleitforschung zur Bundespflegesatzverordnung 1995: Abschlußbericht. Untersuchung im Auftrag des Bundesministeriums für Gesundheit,* Deutsches Krankenhausinstitut, Düsseldorf.

26 Helmig, B. and Lapsley, I. (2001) On the efficiency of public, welfare and private hospitals in Germany over time: a sectoral data envelopment analysis study. *Health Services Management Research,* **14,** 263–274.

27 Institut für das Entgeltsystem im Krankenhaus (2006) Abschlussbericht. Weiterentwicklung des G-DRG-Systems für das Jahr 2007. Klassifikation, Katalog und Bewertungsrelationen. Teil I: Projektbericht, Siegburg, 15 December 2006.

28 Schreyögg, J., Tiemann, O. and Busse, R. (2006) Cost accounting to determine prices: How well do prices reflect costs in the German DRG-System? *Health Care Management Science,* **9,** 269–279.

29 Statistisches Bundesamt (2007) Gesundheitswesen, Fachserie 12 Reihe 6.3 Kostennachweis der Krankenhäuser 2005, Statistisches Bundesamt, Wiesbaden.

30 Simon, M. (1996) Die Umsetzung des GSG im Krankenhausbereich: Auswirkungen der Budgetdeckelung auf die Aufnahme- und Verlegungspraxis von Allgemeinkrankenhäusern. *Zeitschrift für Gesundheitswissenschaften,* **4** (1), 20–40.

31 Gericke, C., Wismar, M. and Busse, R. (2004) *Cost-sharing in the German health care system. Diskussionspapier 2004/4,* Technische Universität Berlin, Fakultät Wirtschaft und Management, Berlin.

32 Grabka, M., Schreyögg, J. and Busse, R. (2005) Die Einführung der Praxisgebühr und ihre Wirkung auf die Zahl der Arztkontakte und die Kontaktfrequenz – eine empirische Analyse, DIW Discussion Papers 506, Berlin.

33 Zok, K. (2005) Das Arzt-Inanspruchnahmeverhalten nach Einführung der. Praxisgebühr. Ergebnisse aus zwei Repräsentativumfragen unter 3000 GKV-Versicherten. *WIdO-monitor. Die Versicherten-Umfrage des Wissenschaftlichen Instituts der AOK,* **2** (2), 1–7.

34 Braun, B., Reiners, H., Rosenwirth, M. and Schlette, S. (2006) *Anreize zur Verhaltenssteuerung im Gesundheitswesen. Effekte bei Versicherten und, Leistungsanbietern.* Chartbook, Bertelsmann Stiftung, Zentrum für Sozialpolitik, Gütersloh, Bremen.

35 Sachverständigenrat zur Begutachtung, der Entwicklung im Gesundheitswesen (2007) Kooperation und Verantwortung. Voraussetzungen einer zielorientierten Gesundheitsversorgung. Gutachten 2007, Bonn.

36 Schneider, M., Hofmann, U. and Köse, A. (2004) *Zuzahlungen im internationalen Vergleich,* BASYS, Augsburg.

37 Reiners, H. (2006) Der Lahnstein Mythos: Die schwere Geburt des RSA, in *Risikostrukturausgleich 2006, Zehn Jahre Kassenwahlfreiheit* (eds D. Göpffarth *et al.*), Asgard-Verlag, Sankt Augustin, pp. 13–34.

38 Schneider, W. (2002) RSA-Reform. Kraftakt mit Augenmaß. *Gesundheit und Gesellschaft Spezial,* **5** (2), 4–7.

39 Busse, R. (2004) Disease management programs in Germany's statutory health insurance system. *Health Affairs,* **23** (3), 56–67.

40 Bundesministerium für Gesundheit (2005) *Statistisches Taschenbuch. Gesundheit 2005,* Bundesministerium für Gesundheit, Berlin.

41 Bundesministerium für Gesundheit und Soziale Sicherung (2005) Gesetzliche Krankenversicherung. Mitglieder, mitversicherte Angehörige, Beitragssätze

und Krankenstand. Monatswerte Januar
bis Juni 2005. (Ergebnisse der GKV-
Statistik KM 1).

42 Bundestags-Drucksache 12/3608 (1992)
Gesetzentwurf der Fraktionen CDU/CSU,
SPD und F.D.P. Entwurf eines Gesetzes
zur Sicherung und Strukturverbesserung
der gesetzlichen Krankenversicherung
(Gesundheits-Strukturgesetz).

43 Verband der Angestellten-Krankenkassen,
e.V. and A.E.V. Arbeiter-Ersatzkassen-
Verband, e.V. (2006) Ausgewählte
Basisdaten des Gesundheitswesens,
Siegburg.

44 Andersen, H.H. & Grabka, M. (2006)
Kassenwechsel in der GKV 1997–2004.
Profile – Trends – Perspektiven, in
*Risikostrukturausgleich 2006 Zehn Jahre
Kassenwahlfreiheit* (eds D. Göpffarth *et al.*),
Asgard-Verlag, Sankt Augustin,
pp. 145–189.

45 Nuscheler, R. and Knaus, T. (2005) Risk
selection in the German public health
insurance system. *Health Economics*, **14**,
1253–1271

46 Höppner, K., Greß, S., Rothgang, H. and
Wasem, J. (2006) Kassenwechsel in der
GKV 1997–2004. Profile – Trends –
Perspektiven, in *Instrumente der Risiko-
selektion – Theorie und Empirie* (eds
D. Göpffarth *et al.*), Asgard-Verlag,
Sankt Augustin, pp. 119–144.

47 Bundesministerium für Gesundheit
(2006) Statistik über Versicherte,
gegliedert nach Status, Alter, Wohnort,
Kassenart 2006. http://www.bmg.bund.
de/cln_041/nn_601096/DE/Statistiken/
Statistiken-Gesundheit/Gesetzliche-
Krankenversicherung/Mitglieder-und-
ersicherte/mitglieder-und-versicherte-
node,param=.html_nnn=true (accessed:
16 July 2007).

48 Stock, S., Redaelli, M. and Lauterbach,
K. (2007) Disease management and health
care reforms in Germany – Does more
competition lead to less solidarity? *Health
Policy*, **80**, 86–96.

49 Wörz, M. and Wismar, M. (2001) Green
politics in Germany: what is Green Health

Care Policy? *International Journal of Health
Services*, **31** (4), 847–867.

50 Greß, S., Focke, A., Hessel, F. and Wasem,
J. (2006) Financial incentives for disease
management programmes and integrated
care in German social health insurance.
Health Policy, **78**, 295–305.

51 Busse, R. and Wörz, M. (2005)
Modernisation of the German Health Care
System, in *Yearbook of European Medical
Law* (ed. M. Sanderfelt), The Institute of
Medical Law, Tallinna Raamatutrükikoda,
pp. 47–55.

52 Bronner, D. (2003) Struktur und Aufgaben
des gemeinsamen Bundesausschusses
nach § 91 Absatz 2 SGB V. *das Krankenhaus*,
95 (11), 866–868.

53 Institute for Quality and Efficiency in
Health Care (2008) Methods for
Assessment of the Relation of Benefits to
Costs in the German Statutory Health Care
System. For Consultation. Version 1.0.
24th January 2008. Download
http://www.iqwig.de/download/08-01-
24_Methods_of_the_Relation_
of_Benefits_to_Costs_Version_1_0.pdf
(accessed: 31 March 2008).

54 Evans, R.G., Barer, M.L. and Hertzman,
C. (1991) The 20-year experiment:
accounting for, explaining, and evaluating
health care cost containment in Canada
and the United States. *Annual Review
of Public Health*, **12**, 481–518.

55 Statistisches Bundesamt (ed.) (1997)
*Datenreport 1997. Zahlen und Fakten über
die Bundesrepublik Deutschland*,
Bundeszentrale für politische Bildung,
Bonn.

56 Statistisches Bundesamt (ed.) (2002)
*Datenreport 2002. Zahlen und Fakten über
die Bundesrepublik Deutschland*,
Bundeszentrale für politische Bildung,
Bonn.

57 Statistisches Bundesamt (ed.) (2006)
Datenreport 2006. *In Zusammenarbeit mit
dem WZB und, ZUMA. Zahlen und Fakten
über die Bundesrepublik Deutschland*,
Bundeszentrale für politische Bildung,
Bonn.

58 Wendt, C. (2007) Sinkt das Vertrauen in Gesunheitssysteme? Eine vergleichende Analyse europäischer Länder. *WSI Mitteilungen*, **60** (7), 380–386.

59 Sachverständigenrat für die Konzertierte Aktion im Gesundheitswesen (2003) *Finanzierung, Nutzerorientierung und Qualität. Band I. Finanzierung und Nutzerorientierung. Band II. Qualität und Versorgungsstrukturen*, Sachverständigenrat für die Konzertierte Aktion im Gesundheitswesen, Bonn.

60 Gerlinger, T., Mosebach, K. and Schmucker, R. (2007) *Wettbewerbssteuerung in der. Gesundheitspolitik. Die Auswirkungen des GKV-WSG auf das Akteurshandeln im Gesundheitswesen*, Diskusionspapier, 2007-1, Institut für Medizinische Soziologie, Frankfurt am Main.

61 Knieps, F. (2007) Der Gesundheitsfonds aus Sicht der Politik, in *Jahrbuch Risikostrukturausgleich 2007* (eds D. Göpffarth *et al.*), Asgard-Verlag, Sankt Augustin, pp. 9–26.

62 Jacobs, K. (2004) Rückblick: Ein System lernt dazu. *Gesundheit und Gesellschaft Spezial. Reform des Risikostrukturausgleichs: Maßarbeit für den Patienten*, 4 (10), 6–7.

6
The Netherlands

Werner Brouwer and Frans Rutten

Population (million)	16.3
GDP per capita (US$ PPP)	34 200
Health spending as % of GDP	9.2
Public health spending as % of total spending	62.5
Health spending per capita (US$ PPP)	3094
Acute care beds per 1000 population	3.1
Practicing physicians per 1000 population	3.7
Life expectancy at birth (years)	79.4
Infant mortality per 1000 live births	4.9

Strategies used

1. Shift from supply-side rationing to demand-side rationing
2. Price and supply side regulation
3. Waiting lists
4. Pharmaceutical sector regulation
5. Demand-side rationing (cost sharing, benefit package limitation)

6.1
Introduction

Worldwide, increasing and virtually uncontrollable health care expenditures appear to be a highly prevalent disease. In part, the 'disease' may be contagious since many countries compare themselves to others in terms of expenditures and coverage, and follow the leaders or the overall trend. Another part, however, seems to be inherent to progress in terms of the wealth of nations, medical technology and notions of solidarity in most Western countries. This leaves most countries fighting the seemingly incurable disease of rising health care expenditures.

Cost Containment and Efficiency in National Health Systems: A Global Comparison
Edited by John Rapoport, Philip Jacobs, and Egon Jonsson
Copyright © 2009 WILEY-VCH Verlag GmbH & Co. KGaA, Weinheim
ISBN: 978-3-527-32110-0

The Dutch situation is not much different from that in other countries. The Dutch health care expenditures are, with some 9.2% of GDP for 2004 (see OECD data, 2007) neither exceptionally low nor exceptionally high. Since early 2000 – when expenditure was 8.0% of GDP, according to the OECD figures – expenditures are however increasing relatively rapidly. This seems to be related with policy changes in the Dutch health care system, that are part of (or at least in line with) a bigger reform of the health insurance and health care system. The Dutch health care system is in a transition from a supply-regulated system with two types of health insurance – private for the richer third and social for the poorer two-thirds – towards a system with uniform private health insurance for all citizens, based on notions of consumer choice and regulated competition. The general idea behind the reform is to improve the efficiency of the health care sector, while maintaining quality, access and solidarity. It may be viewed as the Dutch way of catching the third wave of health system reforms [1], marking a shift in focus as of the late 1990s from pure cost containment to improved efficiency.

A major step in the process was taken on 1 January 2006, when the new Health Insurance Act came into effect and created the legal foundation of the new health care system. The reform has clear consequences for the way the health care sector is regulated, and therefore the way in which costs may be contained. Whereas before 2000 the focus was especially on restricting supply and prices (since the mid 1980s), which was a reasonably successful strategy though not without side effects, the focus is now gradually shifting towards the demand side – compatible with the new policy paradigm and market ordering. This transition, and the challenges it poses to cost containment in the Dutch health care sector, make the Dutch case exceptional and potentially interesting for other countries.

In this chapter we will highlight the Dutch strategies towards cost containment and improving efficiency, drawing on recent publications in this area (e.g. [2–5]). First, in Section 6.2 we will briefly introduce the Dutch health care sector, both its situation before 2006 and after. Then, in Section 6.3 we highlight the most important cost-containment strategies applied in the last decades and observe how cost containment has evolved with the transition, from mostly supply regulation towards regulation of competition. Finally, in Section 6.4 we discuss some future challenges and present some conclusions. Note that this chapter describes the situation, developments and plans in the Dutch health care sector up to 2007.

6.2
The Dutch Health Care System in Transition

Here, we will provide some information about the relevant features of the Dutch health care system, focusing mainly on the second compartment of the Dutch health care sector, that deals with curative health care. In order to facilitate the discussion of the different cost-containment strategies in the following two sections, we will highlight both the Dutch health care system of pre-2006 and the current system. We will also briefly describe the rationale behind the reforms.

6.2.1
The Dutch Health Care System: a Short Description

6.2.1.1 The Origin of the System

The Dutch health care system is best characterized as a Bismarckian system. The *European Observatory* describes this as a "... system of national social security and health insurance introduced into the nineteenth century German empire under the then Chancellor Bismarck. This system is a legally mandatory system for the majority or the whole population to obtain health insurance with a designated (statutory) third-party payer through nonrisk-related contributions, which are kept separate from taxes or other legally mandated payments." (www.europeanobservatory.org)

Considering the history of the Dutch health care system, this description and the German origin need not be surprising. During the Second World War, the German occupier laid the basis for the health insurance system that was in place until 1 January 2006 by introducing the first version of the Sickness Fund Insurance Act. This Act introduced a scheme of compulsory health insurance for the poorer two-thirds of the population. Every person with an income below a certain threshold was therefore obliged to take out social insurance. Those who had an income above that threshold could purchase private health insurance voluntarily, which almost all individuals did. In 1964, the Sickness Fund Act was renewed, and a few years later – in 1968 – a national insurance basis for long-term care was established under the Exceptional Medical Expenses Act (AWBZ). Together, these two acts ensured the financial access to essential care for all Dutch citizens. If people wish to insure themselves against the costs of other medical consumption – that is, of care items not covered under these two acts – they are free to buy supplementary health insurance. Thus, the Dutch health care system comprises three distinct 'compartments', each of which covers specific health care items and has specific insurance arrangements and market-orderings [6]. These are depicted in Figure 6.1 and are discussed in the following section.

6.2.1.2 The Three Compartments

The *first compartment* comprises long-term nursing and care and includes the so-called 'catastrophic' or 'uninsurable' risks. The entitlements to care which fall in the first compartment are embodied in the Exceptional Medical Expenses Act (AWBZ), a

Third compartment
Supplementary Health Insurance
Second compartment
Health Insurance Act
First compartment
Exceptional Medical Expenses Act and Social Support Act

Figure 6.1 The three compartments.

statutory social insurance that covers the entire population against the costs of prolonged nursing and (home) care. It should therefore protect the population against the potentially large financial consequences of serious long-term illnesses or disorders, including the cost of caring for disabled people with severe congenital physical or mental disorders and psychiatric patients requiring long-term nursing and care [6]. Payment is arranged through mandatory (up to a limit) income-dependent premiums. This compartment is highly regulated by the government and administered by regional care offices governed by the major sickness fund in the region, which conclude contracts with health care providers [7]. In the past years, the scheme has been 'modernized', which included the possibility to receive cash rather than traditional in-kind benefits through personal budgets, independent indications for entitlement and functional rather than supplier-labeled descriptions of prescribed care to induce more competition between suppliers of care. More recently, a relatively small part of the entitlements which fell under the Exceptional Medical Expenses Act, especially pertaining to home care and social support, have been removed from the scheme and transferred to a new scheme regulated under the Social Support Act (WMO). This Act makes the municipalities responsible for elements of home care, social support and activating care. Although municipalities are responsible for providing the facilities in these areas of care, the Social Support Act does not specifically stipulate the entitlements. Therefore, differences between municipalities can arise in how specific services are provided, especially as the budget for these provisions is limited and may differ between municipalities [6]. Although the introduction of the Social Support Act was defended by pointing out that municipalities would be better able to address local needs and wants of care recipients, it may be perceived as especially resembling a cost-reduction measure, leaving the municipalities in the difficult position as to how to ration this care (e.g. [2]). In 2006, some entitlements under the Exceptional Medical Expenses Act that could be labeled as having the intention to cure people (rather than to care without the hope or intention of curing) have been shifted to the second compartment, which will be highlighted in detail below. In terms of expenditures, the first compartment comprises some 40% of the total. This percentage is likely to grow further in the future, given the aging of the population, which is typically associated with diseases that require long-term care.

The *third compartment* comprises voluntary supplementary health insurance. This can be freely purchased on a voluntary basis to cover for expenses that are not covered under the first or second compartment, such as specific elements of dental care for adults or homeopathic 'medicine'. Given the nature of the care covered in this compartment, the need for its strict regulation is deemed less necessary by the government. Normally, risk-related premiums can be charged by insurers and there is no legal obligation for insurers to accept individuals who apply for a supplementary health insurance policy. Over 90% of the insured have some form of additional insurance.

The *second compartment* is the one comprising curative care or 'care with a view to cure', as the Ministry labels it [6]; this means insurable, curative care, such as GP visits, specialist visits, pharmaceuticals, hospitalizations, and so on. This compartment is central to what follows in this chapter, and it is in this compartment that the

Dutch health care reforms have mainly taken place. Thus, in the following we will first highlight the situation of this compartment before 2006, and then (in Section 6.2.2) highlight the rationale and plans for, as well as some of the preliminary steps taken towards, the actual reforms that were enforced on 1 January 2006. In Section 6.2.3 we describe in greater detail how the second compartment is currently organized and structured. In order to avoid too much detail and repetition, we describe the systems before and after 2006 stylized, ignoring many of the developments before 2006 that made this large step in the reform feasible.

Up until 2006, the second compartment was operated through two distinct insurance schemes. First, there was the mandatory social insurance scheme, regulated under the Sickness Fund Act. This Act stipulated that Dutch citizens who earned less than some specific income level were obliged to take out social health insurance from a sickness fund. For a long time, these sickness funds acted as regional monopolies, with the insured having no possibility to choose between Sickness Funds but simply be allocated to a regional fund. Later, however, sickness funds were obliged to compete with each other, and the insured were allowed to enroll with the fund which they considered to offer the best deal. It should be noted that this development was an important prelude on the new health insurance system of 2006. In order for individuals to have an incentive to opt for a specific sickness fund, a small nominal premium was introduced (whereas the remainder of the premium was an income-dependent contribution). This premium could differ between different sickness funds, and therefore offered a financial incentive for insured to switch between competing funds. The care covered under the Sickness Fund Act was defined (mostly in nonspecific terms) in the basic benefits package which was centrally set by the Government, and the insured were entitled to benefits in kind. The sickness fund sector was heavily regulated by the Government, both on the insurance side and on the care supply side. An important element in the regulation was the obligation for sickness funds to contract all health care suppliers permitted on the market. This left health insurers with limited possibilities for negotiations, and none for selective contracting, thus seriously undermining their role as the actor responsible for the efficient spending of health care premiums. As was crucial in the development towards a new type of market ordering of the second compartment, this contracting obligation was gradually abolished, providing the sickness funds with more room for negotiations with care suppliers, or even selective contracting.

The second insurance scheme within the second compartment (before the reform) was that of private health insurance. People above the income level specified in the Sickness Fund Act were free to opt for private health insurance. In practice, however, virtually all individuals not insured under the Sickness Fund Act opted for private health insurance, so that around 99% of the population was privately or publicly insured. Private insurance could be purchased from one of the competing private health insurers. Premiums were mostly risk-related and included some mandatory solidarity surcharges which flowed to the sickness fund scheme. The items covered under private health insurance were normally identical to those covered under the Sickness Fund Act (i.e. the basic benefits package roughly applied to all insured). Normally, privately insured were not entitled to benefits in kind, but rather to the reimbursement of health care costs. The private health insurance system used the same health care suppliers as the socially insured – that is, there was no two-tier system in terms of health care supply.

Health care supply was and remains predominantly private. The vast majority of hospitals and other care institutions are privately owned. Since 1983, when hospital budgeting was introduced, hospitals could earn income up to a specified annual maximum, but if this was exceeded the budget of the next year was lowered by the amount of excessive income. Recently, hospital budgets have become more flexible, as will be further highlighted below. Specialists normally act as independent professionals within hospitals and are normally paid on a fee-for-service basis [7], an important exception being specialists in academic hospitals, who are salaried. The fees were first paid by health insurers, but later hospitals received a lump sum to pay their specialists. Traditionally, primary care – especially the general practitioner (GP) – has a strong position and acts as the gatekeeper of the system. Before 2006, GPs were financed in two distinct ways: (i) on the basis of capitation payments for the sickness fund insured; and (ii) on a fee-for-services basis for the privately insured. As of 2006, this distinction is no longer relevant, given the reform of the health insurance system detailed below, and GPs now receive a payment which combines a capitation payment with fee-for-service components.

6.2.2
Rationale for and Steps Towards the New System

6.2.2.1 Reasons for Change

The previous health insurance system of the second compartment was not considered to be optimal for several reasons. One important and obvious point of discontent was the existence of two different health insurance schemes, causing a distribution of contributions to the health care that was perceived to be inequitable. One single scheme could solve that problem. Moreover, there was a general feeling that the incentives for health insurers and health care suppliers should be improved, so as to increase the efficiency of the total health care system. It was especially the discontent with the quality and efficiency of the health care system that pressed the health care system towards a major change.

In order to understand the dissatisfaction with the previous market ordering it is important to note that, during the 1980s and early 1990s, the Dutch government introduced much regulation aimed at controlling total expenditures. This was partly deemed necessary due to the economic recession in that period and the European unification which required a decrease of collective expenditures. While this cost control was relatively successful, it came at a high price. By strictly budgeting hospitals, regulating tariffs, salaries and the numbers of specialists and hospital beds, a tight financial and organizational web was woven around the Dutch health care providers. Improvement of their productivity was not rewarded in this complex system of rules and regulations. Different payment mechanisms and budgets for different care institutions resulted in an impediment of cost-effective substitution, a good continuity of care and created problems such as 'wrong bed-days' (i.e. patients remaining in an expensive hospital longer than medically necessary because of the fact that no beds were available in nursing homes). As De Wolf *et al.* [3] note: "During the 1990s it was widely recognized that the health care system, with its detailed government regulation, was incapable of delivering efficient, sufficient and patient-oriented care".

The strict control of total expenditures and health care supply was accompanied by an almost unrestricted health care demand. Virtually no financial restrictions on health care demand were imposed. This combination resulted in a large gap between demand and supply of care, in terms of quantity as well as in terms of quality. The system increasingly failed to deliver appropriate care in time with waiting lists and considerable waiting times as a consequence. Although the Dutch waiting times could be described as modest from an international viewpoint [8], within the Netherlands the waiting lists were considered an important problem that required solution (e.g. [9]). Removing some of the strict regulations aimed at controlling costs – and therefore hampering increased production and reducing waiting times – was considered necessary.

Another structural problem was the fact that sickness funds were not bearing any financial risk for health care expenditures as these were pooled nationally. This obviously did not stimulate them to act as prudent buyers of care on behalf of their insured and seek ways to actively contribute to efficient care delivery. As Schut *et al.* [10] note: "The primary reason of the reform was to motivate sickness funds to improve the efficiency of health care. The absence of appropriate incentives for sickness funds was perceived as a major problem." Introducing such incentives for health insurers could also translate into an increased pressure on health care suppliers to improve efficiency.

These were major reasons for the Dutch government to engage in substantial health care reform – introducing more competition and thus increasing the efficiency of the health care system. Also, the government wished to share the responsibility for reaching the goals of quality, access, efficiency and affordability of the health care sector with the actors in the health care field, especially the insurers and suppliers of care, since it recognized the complexity (and partly the failure) of central guidance [11].

6.2.2.2 Towards a New System

The direction of the health care reforms meanwhile had become more or less clear and accepted in many relevant decision-making arenas. A very influential and coherent proposal for reform in this context was offered by the Dekker committee [12], which in turn was importantly inspired by the work of Alan Enthoven [13] on consumer choice of health insurance plans. The Dekker committee offered a blueprint for the health care system reform, based on the notion of 'regulated competition' and consumer choice, which has guided subsequent thinking and political discussions to a large extent. This blueprint helped to implement changes over the years that have facilitated a gradual movement towards such a system (see, e.g. [14]). Although the Dekker plans would neither be embraced nor executed completely (an important divergence from the plans, being that the reforms were considered especially appropriate for the second compartment, and not also for the first as proposed by Dekker, e.g. [15]), their influence has been profound.

The principle of a health care system based on regulated competition is straightforward. Competing health insurers purchase health care from competing care suppliers, on behalf of their insured. Health insurers are free to contract only those health suppliers that offer value for money, and to negotiate attractive contracts with those suppliers. This means that health care suppliers have a strong incentive to offer quality care efficiently, as otherwise they risk losing their contracts and therefore clients. The incentives for health insurers to act as prudent and critical buyers of care is enforced by allowing the insured to switch yearly between insurers via a system of annual open enrollment and mandatory acceptance of insured by health insurers. Solidarity (both risk and income solidarity) was ensured by obliging all citizens to take out health insurance for which largely income-related premiums would be levied, as well as a small nominal premium to provide incentives to the insured to select efficient health insurers. This system was believed to combine the best of the market (pressure on all actors to offer value for money) with the best of government regulation (solidarity in premium and access to health insurance and health care). Thus, market forces should make the health care system more competitive, efficient and responsive to demand, while the Government would guarantee and monitor quality and equity.

6.2.3
The Second Compartment as of 2006

The new Health Insurance Act that came into effect in 2006 shows much resemblance with the system advocated by the Dekker committee. Figure 6.2 highlights a number of important aspects of the new scheme, in graphical form.

The insured can indeed opt for the health insurer of their choice on an annual basis. Enrollment is open and people cannot be rejected by a health insurer for basic

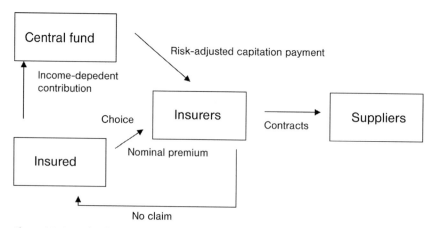

Figure 6.2 A graphical representation of important features of the second compartment.

health insurance. The health insurers are responsible for contracting sufficient care for their insured, and largely for the financial risk they run. The insurers have two forms of income. First, they receive a nominal premium directly from the insured. Unlike the Dekker proposal, this premium is substantial, covering about half of the total contribution (while original plans mentioned percentages of between 10 and 20%). This nominal premium cannot be differentiated between the insured choosing the same insurance policy. The other half of the contribution is levied through an income-dependent contribution, mostly paid by (and compensated by) employers. This income-dependent contribution is collected in a central fund. Subsequently, insurers receive a risk-adjusted payment from this central fund on the basis of the risk-profile of their insured. The current risk-equalization scheme is the most sophisticated in the world, with risk adjusters such as age, gender, labor status, region, pharmaceutical cost groups and diagnostic cost groups. The pharmaceutical cost groups and diagnostic cost groups are proxies for health status based on the use of pharmaceuticals and past hospital use, respectively, and importantly improve the predictive power of the risk-equalization scheme (e.g. [16–19]). This scheme should ensure that health insurers are adequately compensated for the financial risk they bear, given that they are obliged to accept all individuals and cannot charge different nominal premiums for people with different risk profiles. When the risk-equalization scheme corrected perfectly, the insurers would have no incentives for *cream skimming* – that is, selecting good risks – and would all operate on a level playing field. The nominal premium of an insurer would then signal its efficiency to the insured. In spite of the level of sophistication of the current risk-equalization scheme, it still is not a perfect corrector (e.g. [20]), which implies that the nominal premium may still signal the risk profile of the insured rather than the efficiency of a health insurer, and that incentives for risk selection continue to exist.

An interesting feature of the new scheme, especially in the context of this chapter, is the no claim rebate (which was introduced already in 2005 for the sickness fund scheme). When an insured uses less than €255 worth of care in one particular calendar year, he or she will receive a refund of €255 minus the incurred costs in April of the next year. This should make people more prudent users of care. This scheme is discussed in more detail in the next section, and signals an important change in terms of cost containment. Whereas, the emphasis used to be placed on supply-side regulation, it is now shifting towards demand-side regulation, as the former does not mix well with a system of regulated competition. This has important implications for cost-containment strategies, which will be highlighted in the next section.

6.3
Cost Containment Through the Years

In this section we highlight the main cost-containment strategies that have been used in the Dutch context over the past few decades. From the transition of the Dutch health care system as described above, it may be clear that before the reform such strategies focused on the supply side, whereas after the reform they were targeted more at restraining demand. Here, we focus on the main cost-containment strategies introduced before the reform, and how the situation evolved in the period before and since the reforms. In Section 6.3.1 we highlight price regulation, budgeting and supply restriction, focusing mainly on the hospital sector, while in Section 6.3.2 we highlight rationing through waiting lists. Section 6.3.3 specifically examines the regulation of the pharmaceutical market.

6.3.1
Price Regulation, Budgeting and Supply Restrictions

6.3.1.1 Supply-Side Regulation
Cost containment in the previous health care system was predominantly based on supply restrictions, price regulations and budgeting, since in most health care systems – as was the case in the Netherlands before the reform – competition between providers is often nonexistent (or is at least very limited), and prices are not automatically controlled through market mechanisms. Moreover, given that many health care systems often impose measures to keep the supply of health care relatively low (e.g. through physical restrictions such as the number of hospital beds or the number of physicians, or through financial restrictions such as closed-ended hospital budgets), suppliers have considerable market power and price regulation is often required in such systems to keep prices at an acceptable level. The situation in the Netherlands was no different in that respect. Hospital Planning was an important government responsibility (Hospital Facilities Act with special licensing of advanced facilities in article 18 of this Act) and the Health Tariffs Act from 1980 formed the legal basis for price regulation in the Netherlands. The latter Act was executed by a central body (the Central Organization for Health Care Tariffs) which set the prices for hospital services and for other health care providers. In 1983, in order to be better able to contain hospital expenditures, strict, closed-ended hospital budgeting was intro- duced. Initially, the budgeting system was very simple and crude as it was based on historical costs and an allowable percentage increase for inflation and autonomous growth. However, this was deemed unfair and inefficient as it basically rewarded formerly inefficient hospitals and effectively punished their already efficient counter- parts. Gradually, therefore, the system was adjusted and became more sophisticated, using parameters such as catchment area, number and type of specialized functions (e.g. open-heart surgery) and a number of production parameters. In that way, those hospitals serving a large population, offering specialized functions, or performing well in terms of parameters such as outpatient visits, numbers of patient-days, and so on, could be financially rewarded. Nonetheless, the hospital budget was closed-ended and binding; the budgets of the hospitals were set using the formula, and subse- quently they could earn up to the budget limit through billing various health insurers on the basis of their activities. If hospitals billed more than their predefined budget in a particular year, this surplus was deducted from the budget in the following year. The costs incurred by the hospital should obviously stay below its earnings, and therefore below the budget limit. Otherwise losses could occur which, in the extreme case,

could lead to the bankruptcy of private hospitals. Thus, the (financial) incentives for the hospitals were clear: bill the insurers up to the maximum amount and keep the costs as low as possible at the same time. Incentives to increase productivity or for delivering excellent quality care or innovative were lacking, therefore.

6.3.1.2 Towards a Demand-Driven System

Obviously, in light of the new health care system, these incentives needed to be redirected. The lack of incentives for productivity and (dynamic) efficiency were part of the reasons for the health care reforms to be started. In fact, according to Alain Enthoven [21], it is especially the increased efficiency on the health care delivery market (in contrast to the health insurance market) that should result in large gains, since it is there that the majority of the costs of the health care sector are incurred. In order to provide hospitals with incentives for productivity and efficiency, the old budgeting regime needed to be abandoned and a new way of financing hospitals and care providers developed and implemented. The introduction of diagnosis– treatment combinations (DTC)s, labelled DBCs in the Netherlands, (the Dutch version of diagnosis-related groups (DRGs)) which creates a typology of hospital products by looking at both diagnosis and the applied treatment, marked an important step in that process. By defining hospital 'products' in terms of DBCs it is possible to price them individually. Ideally, such prices are formed on a competitive market through negotiations between health insurers and hospitals. However, in the absence of an adequately functioning market (because of too much market power or other forms of market failure) an (Government) agency could also set prices for these products. Currently, both ways of price-forming are applied in the Netherlands; the prices of all DBCs on the so-called 'list A' are currently fixed by the Dutch Health Care Authority (NZA, which now incorporates the 'old' Central Organization for Health Care Tariffs). The prices of the DBCs included in the so-called 'list B' (covering nonacute treatments), however, can be set in the negotiations between hospitals and health insurers (for further details, see Ref. [22]). The fact that the DBCs on list B currently account for some 10% of the total hospital expenses indicates that the role of free negotiations on the price of hospital care is still limited. The Dutch government has decided that negotiations on prices for (certain parts of) the DBCs in the A segment is not due until sufficient experience is gained with the free pricing in the B segment. A careful and deliberate increase of the scope for free price negotiation is foreseen in which a first step would be to increase the free negotiable segment to 20% of total hospital expenses [23].

In order to increase the pressure on other prices in the health care sector, the Government has announced the intention of introducing yardstick competition as of 2010. The Dutch Health Care Authority [23] indicates that, for more than 50% of elective treatments, free prices under a yardstick will be introduced. The yardstick will use averages of relevant DBC prices, but the proposed method allows higher prices for specific DBCs for hospitals as long as the total revenues of a hospital remain below the sum of all yardstick prices for the DBCs relevant for that hospital. This cap on earnings creates a new budget which currently appears to be used to act as a ('old fashioned') cost-containment measure. Moreover, the Health Care Authority suggests the gradual introduction of this system of yardstick competition. In

2012, yardstick competition should be completely replaced with free pricing for elective treatments [23]. Hence, the NZA basically proposes to divide the hospital sector in the coming years into three distinct parts: one with fixed prices (30%; intensive care, top clinical treatment, expensive medication); one with free pricing (20%); and one under yardstick competition (50%). Note that these are the plans as formulated in 2007, which indicate the general direction of health policy, yet may be (expected to be) subject to changes.

In terms of other supply regulations, the Government has also gradually released its grip on the health care market somewhat. The Hospital Facilities Act was substituted in 2006 by the Act on the Admission of Health Care Institutions. Now, new entrants can come onto the market more easily and the regime for building care facilities has become less restrictive. This is of course crucial in the context of regulated competition, as a lack of supply results in market power for suppliers and a weak position for health insurers in the negotiations with suppliers.

In summary, both price and supply regulation were used successfully in the Netherlands to contain costs, and low growth rates of health care expenditures during the years after the introduction of these measures were the result. However, it gradually became clear that success in terms of cost containment through supply regulation had adverse consequences, such as a lack of quality, efficiency and innovation. Therefore, there is a slow – but clear – movement away from these instruments of cost containment. New ways of regulating the market, for example through the introduction of yardstick competition, which has been criticized often in the literature (e.g. [24]), are meant to gradually work towards full competition. Meanwhile, it should be recognized that such temporary measures may create their own dynamics; for example, yardstick competition does not sufficiently take into account variations in quality and therefore may result in adverse incentives for suppliers and health insurers. Moreover, it should also be recognized that abandoning the supply-side restrictions will ultimately require new ways of cost containment, which focus more on the demand side.

6.3.2
Waiting Lists

Limiting the supply of care through cost-containment measures as discussed above while not restricting demand, normally results in large discrepancies between demand and supply. The consequence of this was the emergence of considerable waiting lists for different procedures in the Dutch health care sector. While waiting times in the Netherlands were relatively short when compared internationally, they were perceived as problematic by Dutch standards. To illustrate the latter point, Table 6.1 presents the results from the Health Interview Survey 1993/1995 performed by the Dutch CBS [26] as presented in Brouwer and Schut [9].

The data in Table 6.1 indicate that, on average, patients had to wait 58 days before they received the treatment they required, and that they felt this clearly exceeded the acceptable maximum waiting time (by a factor 2.5 on average).

Interestingly, however, these waiting times need not only be considered as a *result* of cost-containment strategies, they can also be seen as a (deliberate)

Table 6.1 Waiting time in the Netherlands; real and acceptable [9, 26].

Specialist	Average waiting time (days)	Acceptable maximum waiting time (days, as indicated by patients)
Gynecologist	51.8	41.7
Orthopedist	66.5	23.0
Eye doctor	88.5	30.6
Neurologist	40.1	11.2
Psychiatrist	57.4	23.4
ENT-specialist	50.6	28.9
General surgeon	59.6	23.8
Plastic surgeon	140.1	55.5
Total	58.3	24.2

ENT = ear, nose and throat.

cost-containment strategy in itself. The existence of waiting times can effectively reduce the demand for care. Lindsay [27], for instance, indicates that demand for health care depends (also) on the waiting time for health care, which may be perceived as a price of health care. Economic theory also suggests that normally when the (waiting) price of health care is high, demand is low, and *vice versa* [28]. Siciliani and Hurst [8] recently presented a generic model of waiting times and indicated that these, of course, depend on the interaction between demand and supply. This also indicates that it may be difficult to reduce waiting times. After an anticipated reduction of waiting time, it will rise again due to increased demand, which has been compared (see Ref. [29]) with the act of digging in the sand on the beach – the sand keeps filling the hole while you are digging.

It is interesting to see how this interaction between demand and supply in health care exactly behaves. Martin and Smith [30], for example, indicate on the basis of 4000 observations in a cross-regional study in the British NHS that waiting times are low when the number of available hospital beds (a proxy for supply) is high. Furthermore, demand is higher when supply is higher (negative elasticity of the price of waiting), but this increase in demand is relatively small, so that "... increased NHS resources can bring about reductions in waiting times" while the "... associated stimulation of demand is relatively trivial" according to Martin and Smith [30]. In the Dutch case, after some time the strategy of allowing and rewarding more production by hospitals helped to bring down waiting times. Currently, waiting times are no longer an important health policy issue.

6.3.3
Policies to Contain Pharmaceutical Expenditure

The pharmaceutical market in the Netherlands has traditionally been quite liberal, and was not much regulated before 1991 [31]. Producers were free to set their prices and consumers normally did not face any copayments for pharmaceuticals. However, as the annual growth in pharmaceutical spending began to increase and exceeded 8%

in for instance 1987 and 1990, the concern about the consequences of this liberal system grew rapidly. This resulted in a number of measures to contain pharmaceutical expenditures during the early 1990s. We will highlight here the restrictions on reimbursement and price, and the incentives to prescribers and pharmacists.

6.3.3.1 Introducing Reference Pricing

A first response to the rising pharmaceutical expenditures was the introduction of the Drug Reimbursement System (DRS) in 1991.[1] The DRS is a so-called 'reference price system', which pertains solely to the reimbursement of extramural pharmaceuticals bought in the city pharmacy. When a new medicine is considered for incorporation into the DRS, and is a therapeutic equivalent of (an) already existing medicine(s), it can be easily added to one of the existing clusters of the DRS (the so-called list 1A), which group similar drugs with similar pharmacotherapeutic effects. As Stolk and Rutten [22] indicate: "Medicines that have a similar area of application and a comparable method of administration, with no clinically relevant differences in their properties and intended for the same agreed group together make up a cluster in schedule 1A."

Each of these clusters has a reimbursement limit, which was initially set at the average of the prices of the drugs in the cluster. Adding a new medicine to the cluster will not increase costs, as it is used in the same target group as the existing drugs and cannot be priced at a higher level (at least not if the full price is to be reimbursed [22]). If the price of a drug is higher than this limit, the difference between the limit price and the actual price needs to be paid out-of-pocket by the patient. If the manufacturer claims that their product is not therapeutically substitutable but in fact more effective than current medication, they may ask for a premium price and then has to apply for admission on the so-called list 1B. In this application, the manufacturer must demonstrate the superiority of the product. This list contains the nonclusterable drugs, for which no reimbursement limit exists.

As Rutten [31] explains, shortly after the introduction of the DRS, a convergence of prices towards the reimbursement level could be observed in order to avoid copayment by patients and losing market share. This resulted in a small average reduction in prices. Soon after, however, the expenditure went up again. This increase was mainly caused by the introduction of highly priced new compounds that could not be clustered with existing drugs. This led the government to impose drastic measures and to not reimburse new innovative drugs between 1993 and 1999, except when such drugs offered the first pharmacological option for a previously intractable condition [31]. As opposition to this restrictive policy was mounting, the Government decided gradually to introduce value-based reimbursement with a phasing in from 2002 onwards. Since 2005, companies are obliged to submit evidence about the cost-effectiveness of the medicine if they want a premium price and be placed on list

---

1) De Wolf et al. [3] note that the Medicines Price Act, introduced in 1996, is a second Act aimed at containing pharmaceutical prices. It limits Dutch prices to the average price of the same drug in Belgium, France, United Kingdom and Germany. The maximum price allowed on the basis of the Medicine Price Act may well be pushed downward by the DRS.

1B [22]. Although this new form of value-based reimbursement is clearly aimed at promoting efficiency rather than cost containment, the practice is that the criterion 'budget impact' still receives much weight in the final decision making by the Minister. There are some examples where cost-effective products were not admitted to the DRS because of the high budget impact that was predicted (e.g. the case of sildenafil [32]).

6.3.3.2 Prescribing More Generics

As De Wolf *et al.* [3] indicate, the DRS appears to have failed as a real cost-containment strategy. Rutten [31] states: "So we have seen only a short-term and fairly small effect of the reference price system and no long-term benefits as has also been observed in other countries with price reference systems." Not only did it contribute little to decreasing the costs of clustered drugs, the DRS also offered no incentives to lower prices below the limit. Especially with the increased attention for generic prescribing in the Netherlands, this led to a situation in which the prices of generics increased rather than decreased, leading to even more pharmaceutical expenditure. To understand the mechanism behind this development, the position of the pharmacist in the chain of prescribing and dispensing pharmaceuticals needs to be highlighted. Besides checking the prescription, informing the patient and ensuring that the patient does not receive different pharmaceuticals which interact, the pharmacist can also substitute between brand products and generics (but only if the GP prescribes generically). Moreover, the pharmacist buys pharmaceuticals from wholesalers and generic producers. When a GP prescribes generically, the pharmacist may therefore deliver any generic or brand product (as long as it suffices the prescription). The price that the pharmacist can claim with the insurer when he or she delivers a product is the so-called 'list price'. This price needs to be at or below the DRS limit of the relevant cluster. But what happens if a generic producer sets the list price just below the cluster limit while production costs are very low? In that case, the producer can offer the pharmacists (extremely) large discounts if he or she were to deliver this particular generic product. In other words, a system of margin-competition resulted that was made possible by the DRS system. Thus, generic products were priced relatively high (an average price difference of 4% was reported between generics and specialties) and this led to high expenditures for pharmaceuticals and an upward pressure on insurance premiums as well as on profits for pharmacists. As Rutten [31] notes: "... most of the benefits of prescribing more generic products have fallen to the pharmacist in stead of to the insurer or, in the end, the patient or insured".

6.3.3.3 Role of Health Insurers

In light of the new health care system, it was hoped and advocated [33, 34] that health insurers would end this undesirable situation. However, they were unable to find effective strategies to do so, reflecting the strong position of Dutch pharmacists. Therefore, the previous Minister of Health forced the acceptance of a 'voluntary agreement' (covenant) upon all parties involved, that a 40% reduction in generic prices would be effected. This brought down pharmaceutical expenditures considerably. Recently, it was announced that there are plans to bring down the prices of pharmaceuticals even further (thus lowering the profits through discounts for the pharmacists). In return, the market for pharmaceuticals will not be liberalized further for some time. While very effective as cost-containment strategy, such measures obviously are not compatible with the general idea behind the new health care system of more competition, a larger role for the health insurers, and less government regulation.

Besides measures regarding reimbursement and pricing there were attempts to stimulate cost-consciousness among prescribers. De Wolf *et al.* [3] indicate that GPs are generally free to prescribe whatever medication they see fit for the patient at hand, without having to consider the cost-consequences of their choice or incentives to be prudent prescribers. They indicate:

> "There are guidelines for some disease areas, in the construction of which cost-effectiveness has played a role and there have been attempts to influence GPs through an electronic prescription system. However, the GPs often have limited direct incentives to strictly follow guidelines (which by the way normally do not specify the specific brand of pharmaceuticals needed but offer more general guidance) or to work with the electronic prescription system (EPS). The results of experiments with the EPS were rather disappointing in terms of cost savings. Although 70% of GPs reported that they use the system, the estimated savings on drug costs appear to be modest: €7–15 million as compared to the expected €139 million reduction in costs [35]. The evaluation concludes that the EPS may have a more positive impact on the quality of drug prescriptions than on the cost. Guidelines and the EPS may also be perceived as a limitation on the professional autonomy of medical doctors."

More recently, (some) health insurers are becoming more active in this area. One health insurer offers GPs – on a voluntary basis – the possibility to earn a bonus if they, while conforming with professional guidelines, would prescribe the least-expensive drugs available. While this has been contested several times in Court, this strategy was approved and seen as being in line with the rationale for the new health care system. This can be an important development in the influence of health insurers in prescribing behavior.

6.3.3.4 Hospital Drugs

As noted, the DRS only affects the reimbursement of extramural pharmaceuticals. However, pharmaceuticals in the intramural setting are also responsible for high (and rapidly increasing) expenditures. These pharmaceuticals do not constitute such a large problem in terms of cost containment, as they traditionally fall under the budget of the hospital. However, hospitals increasingly report difficulties in financing these increasingly more expensive pharmaceuticals, and this has resulted in quite some variation in the degree to which patients have access to these expensive medicines. This has led to considerable (media and political) attention in recent years. In order to reduce these problems, the Government decided to allow health insurers to reimburse 80% of the costs of a specific list of very expensive hospital drugs, leaving only 20% to be paid out of the hospital budget. It was decided however, that such a regime would be conditional on the monitoring of the performance (i.e. cost-effectiveness) of these expensive drugs in practice. Continuation of the additional financing depends on the results of this monitoring after three years. This development can therefore be characterized as a form of value-based reimbursement, but now evidence about the performance in daily practice is emphasized rather than predictions on performance based on well-controlled experiments. Hospitals are likely to have more freedom regarding earnings and spending in the near future, and their microefficiency will be stimulated through the further use of the DBC system described above, which is intended also to include drug costs in the future. New developments in terms of new (expensive) drugs should then be automatically absorbed into the price per DBC.

We therefore see an obvious trend in the Netherlands towards a more value-based reimbursement of pharmaceuticals, with an important role for cost-effectiveness information. However, while this may be the trend, this trend (again) is frequently disturbed by more short-term cost-containment strategies of the Government, which are not always in line with the new market ordering of the health care sector.

6.3.4
Demand Reduction

As indicated above, a system of regulated competition does not mix well with supply-side restrictions. Basically, two main options for cost containment exist that are compatible with the new market ordering of the health care system: (i) cost-sharing; and (ii) limiting the basic benefits package. Both are aimed at the demand-side rather than the supply side.

6.3.4.1 Traditional Ways of Cost-Sharing
Cost-sharing has long been a controversial issue in the Dutch health care sector, against which there appeared to be quite some political and societal resistance, especially in the light of their possible adverse distributional consequences [4]. The Dutch health care system is paid, for a relatively small part, via copayments (some 8.5% in 2004), this proportion being especially low in the curative sector. Some attempts to introduce 'own payments' have been made over the years, but normally these were designed in such as way as to avoid any adverse distributional consequences, and therefore failed to be effective in reducing care consumption. They were often abolished soon after their implementation, as administrative costs tended to exceed any savings from a reduction of consumption. Good examples of this are the introduction of a small mandatory deductible in the former Sickness Fund Insurance of about €90, and the introduction of a fixed own payment for medicines. As Van de Ven [36] explains, the latter incentive was designed in such a way that a fixed copayment of about €1.15 per prescription needed to be paid by the insured to the pharmacist. The consequence was that the number of prescriptions went down, but that per prescription the amount of pharmaceuticals went up: in other words, the copayment per pill obviously went down but more medicines were wasted. Van de Ven also indicates that many studies examining the effects of

copayments are hampered by methodological problems such as selection bias (e.g. when studying the effects of voluntarily chosen deductibles) and multiple changes (more than one policy change in the study period, making it impossible to specify what caused an observed effect). Still, a number of good reports have been made on the effects of different types of copayment in the Dutch situation. For example, Rutten [37] studied the price sensitivity of socially insured women when faced with a copayment of 25% for an inpatient day to deliver their babies in the hospital. He reported a price elasticity of −1.5, indicating that increasing the price of an inpatient day with 10% would result in a decrease in the number of hospital days of 15%. Van Vliet [38], moreover, reported on the effects of deductibles, indicating that they have a clear (albeit limited) negative effect on health care expenditures. While Van Vliet reports an overall price elasticity of −0.079, there were important differences between the different types of care. For physiotherapy and GP visits the price sensitivity was relatively high, but it was relatively low for specialist visits and drugs. The demand for hospital care even appeared to be insensitive to prices.

6.3.4.2 The No-Claim Rebate

In the current system a no-claim rebate is in effect. This is an instrument aimed at increasing the cost-awareness of care consumers and reducing consumer moral hazard by limiting unnecessary care consumption through financial incentives ([39] J. Holland et al., unpublished results). The design of the rebate is simple: individuals who consume less than €255 worth of care in a particular calendar year are entitled to a rebate of €255 minus the incurred health care costs. As long as the difference between these two is positive, a refund will be received. So, if a person consumes €124 worth of care in 2007, he or she will receive a refund of €131 in April 2008. This no-claim rebate is a mandatory scheme and was introduced in January 2005 for the sickness fund-insured. When the health insurance reform had taken effect in January 2006, the no-claim scheme was extended to cover the entire Dutch population. In order to finance the scheme, the nominal insurance premium is increased for all insured by some €90 a year. The instrument was seen as a relatively friendly way of reducing moral hazard, as it would not create a real financial threshold to consumer care. Moreover, the administrative burden related to the instrument was expected to be less than for a 'normal' copayment. In order to increase its friendliness and to reduce opposition against the instrument (especially by GPs), it was decided that neither GP visits nor maternity care would fall under the scheme. Therefore, if patients visited the GP, their refund would remain intact; only if the GP were to refer them to a specialist or prescribe care (e.g. medication) would the related costs be subtracted from their no-claim refund.

Given its nontraditional design, it was unclear beforehand what to expect of the no-claim scheme in terms of its ability to reduce care consumption. For the Government, a first and certain *financial* effect from the no-claim scheme was the shift of €1.4 billion (the total amount of money transferred within the no-claim system) from public to private expenditure, thus reducing public expenditures.

However, this is an administrative rather than a real success and does not reduce the total health care expenditures. The latter could only be realized through a behavioral effect. The Dutch Bureau for Economic Policy Analysis [40] estimated this behavioral effect of the no-claim scheme at an annual cost reduction of €200 million, assuming however that the behavioral effect of the scheme was equivalent to that of a deductible of similar size. This assumption was contested, however. Schut [41] for example pointed out that the design of the no-claim system results in a rather weak link between care consumption and the delayed financial reward. Moreover, rewards like the no-claim rebate induce a weaker behavioral effect than do losses (deductibles) [42]. In addition, for important groups of care consumers the no-claim scheme does not provide any incentive whatsoever, as the annual care consumption of many chronically ill will certainly exceed €255. All of these considerations led to the expectation that the no-claim rebate had little potential for curbing health expenditures [41, 43]. Given its design, one might even wonder whether a behavioral effect due to the no-claim scheme would be desirable. As GP visits are excluded from the scheme (in order to maintain unrestricted access to professional care) and GPs are still the 'gatekeepers' of the Dutch health care system, any subsequent care consumption is prescribed by the GP and therefore has a greater probability of being medically necessary. Meanwhile, several studies have investigated the effect of the no-claim rebate scheme since its implementation, and indeed it does not appear to have any significant effect on the behavior of care consumers [39]. Therefore, as of 2008, the no-claim rebate will be abolished and replaced by a mandatory deductible of €150.

In conclusion, the Dutch experience with copayment schemes has been very disappointing so far. In most cases, the administrative burden of implementing these copayment schemes were not compensated by their benefits. In all cases the schemes were politically motivated rather than based on sound scientific evidence, and have led to substantial wasted resources. Perhaps the increasing pressure to impose demand-limiting measures will induce firmer action and better results in the future.

6.3.5
Limiting the Basic Benefits Package

6.3.5.1 Criteria for Defining the Basic Package
As indicated above, another way of limiting demand is to limit the entitlements of people through narrowing the basic benefits package. This policy option has been adopted more frequently during the past few years, but it remained largely unclear on what basis the decisions to remove certain entitlements (e.g. elements of dental care and physiotherapy) were made. Those decisions often appeared rather opportunistic, which was disappointing as there is a long history in the Netherlands in terms of considering the criteria for delineating the basic benefits package. One very influential report in this context was that of the Committee on Choices in Health Care (or the Dunning Committee) from 1991 [44]. This

committee proposed four explicit criteria to guide decisions regarding the basic benefits package:

- Necessity: does the health problem make an intervention necessary?
- Effectiveness: does the intervention achieve a health improvement?
- Efficiency: does the intervention achieve the health improvement at reasonable costs; does it offer value for money?
- Own responsibility and payment: need we pay for this intervention collectively, or can we leave it to individual responsibility?

The committee depicted the decision-making process as a funnel (the Funnel of Dunning, as it is normally known): interventions need to pass all four sieves (criteria) in the funnel in order to fall into the basic benefits package. Especially, the first three criteria have been used in the further operationalization of the funnel, and are frequently mentioned in the context of what should be covered under the Health Insurance Act. For instance, the Department of Health [45] indicates that the basic benefits package should contain care which is "... is checked against its demonstrable effect, cost effectiveness and the need for public financing". But, as a high-level Dutch Advisory Board (the so-called Raad van State) indicates, there is no evidence that the Government effectively and systematically evaluates care on the basis of these criteria. Nor is it clear for that matter that the care currently incorporated in the basic benefits package (is used in a way that) satisfies these criteria.

6.3.5.2 Cost-Effectiveness as an Important Criterion

As Rutten and Brouwer [46] indicate, systematic attention to the cost-effectiveness of (delivered and insured) care is lacking, even though there is a broad consensus on the way in which care should be judged. For extramural pharmaceuticals the situation is relatively favorable, since as of 2005 it is mandatory for pharmaceutical companies to provide a pharmacoeconomic evaluation of a drug when applying for reimbursement at a premium price. Nonetheless, the cost-effectiveness of these drugs is not tested in real life, nor are there normally any medical guidelines that are systematically based on the criteria indicated above. The situation is (much) worse for most other health technologies (e.g. assistive devices and diagnostic equipment), where economic evaluations occur at best on an *ad hoc* basis. Often, new technologies are simply introduced at the level of specialists and hospitals, and when they become 'usual care' they are reimbursed. The plea for a better and more systematic evaluation of health care in this context is gaining attention, however. At present, a dominant operation of the first three criteria of the funnel of Dunning is increasingly accepted, and there is strong pressure to use these criteria for all types of care covered in the basic benefits package. For example, a recent report of the Dutch Council for Public Health and Health Care [47] entitled 'Sensible and sustainable care' stresses this and argues that so far "... decisions regarding payment or nonpayment for medical treatment are only based to a limited degree on 'hard' factors such as cost-effectiveness, and much more on less transparent considerations as a result of pressure by lobby groups like consumer organizations, the media, and so on. This means that limits are indeed being set at present, but on an ad hoc and somewhat random basis. The result is that the available resources are not being deployed as efficiently as possible."

6.3.5.3 Combining Efficiency and Equity

The Dutch Council for Public Health and Health Care proposes the use of a model of evaluating interventions which was developed by the institute for Medical Technology Assessment (iMTA) [48, 49]. In this model, the effectiveness and cost-effectiveness is assessed by means of a cost–utility analysis, with quality-adjusted life-years (QALYs) as the outcome measure. The idea behind the model is that the results of a cost–utility analysis (the costs per QALY gained ratio) should be judged in the context of the necessity of treatment. The more necessary the treatment is, the higher the costs per QALY can be yet still be considered acceptable. Necessity in this respect is calculated as the proportion of health lost due to an illness when not treated. The larger this proportion, the higher the necessity score. Figure 6.3 depicts the acceptability of different costs per QALY in relation to the necessity of care.

The data in Figure 6.3 indicate that the threshold for costs per QALY therefore is not a fixed value (as currently often thought and depicted as the flat line in the figure), but rather increases with severity (the curved line). In other words, society is willing to pay more for a given health improvement when the underlying disease is more severe. One important uncertainty in the model – as in other countries – is the exact location of the line. What is society willing to pay for some health improvement in the different disease categories? The Dutch Council for Public Health and Health Care [47] recently proposed a maximum threshold (for the most severe diseases) of €80 000 per QALY gained. Moreover, it argued that treatments for very mild diseases should not be eligible for funding.

It appears that thoughts about, and the use of, cost-effectiveness analyses in the context of delineating the basic benefits package in the Netherlands are developing and, by using these criteria in defining the package, cost containment may be achieved through a more strict evaluation of health technologies (e.g. lower thresholds). Until now, however, the removal of entitlements from the basic benefits

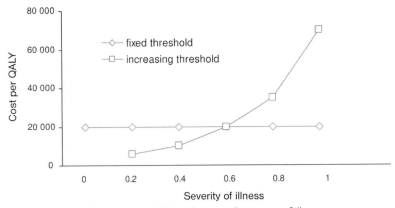

Figure 6.3 Acceptable costs per QALY in relation to the severity of illness.

package has not been based on such clear criteria. The Dutch Health Insurance Board (CVZ) is the main advisor of the Minister regarding delineation of the basic benefits package, and currently is increasing the use of economic evaluations.

6.4
Future Challenges

In this chapter we have described some past experiences and current trends in cost containment in a rapidly changing Dutch health care sector. Some of the important measures that have been imposed over the years are highlighted in Table 6.2.

The main trends in health care policy in general – and that of cost containment in particular – is that, after the introduction of measures aimed at cost containment via regulation of supply and prices, there is a clear tendency towards more focus on efficiency, competition and compatible cost containment via limiting demand. Therefore, the cost-containment policies must be considered in the context of

Table 6.2 The important measures introduced into the Dutch health care sector over recent years.

Year	Measure
1981	Introduction of Health Care Prices Act, enabling the limiting and fixing of prices in the health care sector
1983	Introduction of hospital budgets (further refined in later years), limiting the total revenues of hospitals to a maximum
1991	Introduction of the Drug Reimbursement System, allowing prices of extramural drugs to be limited
1996	Introduction of the Drug Prices Act, which limits the prices of pharmaceuticals to an average of these drugs in Belgium, France, Germany and the United Kingdom
1997	Introduction of own-payment sickness fund-insured of some €90 (200 guilders). Abolished as of 1999 due to limited effectiveness and high administrative costs
2001	Government allowed hospitals to earn more income above the budget to allow more production (in an attempt to reduce waiting times)
2004	First agreement to lower prices of generic drugs with some 40%, was prolonged later
2005	Introduction of no-claim rebate scheme for sickness fund-insured (after 2006 for all insured)
2005	Economic evaluation required for innovative new drugs requesting reimbursement
2005	Negotiations between health insurers and hospitals on the price of Diagnosis–Treatment Combinations (DTCs) for 10% of hospital care
2006	Introduction of new Health Insurance Act, introducing one mandatory basic health insurance for all Dutch citizens

changing the market ordering of the health care system to one based on regulated competition and consumer choice. Given that limiting demand is uncommon and unpopular in the Dutch context, while important steps have been taken in terms of lifting supply-side restrictions, the tension between the health care goals of efficiency and affordability (i.e. cost containment) appears to be growing. Therefore, while the principal changes in the health care sector can be seen as steps towards the introduction of regulated competition, the wish to contain costs sometimes results in interventions that are not in line with this reform. Important examples of such interventions are the caps placed on the budgets of hospitals, the introduction of yardstick competition, and the new collective agreement on pharmaceuticals which basically trades the liberalization of that market for a short-term profit skimming of the excess gains of pharmacists. It appears difficult to forego such short-term successes in the area of cost containment, even when the long-term vision for the health care sector is compromised by them. Moreover, it appears difficult to let go of some of the traditional measures to control the health care sector. Finding a balance between the goals of efficiency and affordability, as well as finding the political courage to move forward, remains an important challenge.

Regulation in the Dutch health care sector is indeed aimed at improving efficiency rather than at cost containment. Regulated competition is not compatible with the 'old' cost-containment strategies aimed at the supply side. Given the limited possibilities to limit demand directly (i.e. introducing cost sharing or limiting the basic benefits package) and the limited experience in that area in the Dutch context, this is by no means an easy transition. Much effort is required to start investing in instruments that will enable patients and insured to become prudent buyers. Introducing serious copayments and limiting entitlements on the basis of the criteria of necessity, effectiveness and cost-effectiveness should be part of a strategy to achieve this. In this way, the Dutch health reform may reach full implementation and catch the 'third wave' of health reforms [1] while containing costs to ensure affordability.

References

1 Cutler, D. (2002) Equality, efficiency, and market fundamentals: the dynamics of international medical-care reform. *Journal of Economic Literature*, **60**, 881–906.

2 Rutten, F.F.H. and Brouwer, W.B.F. (2003) The price of demand-driven care. In *Zorgaanbod en Cliëntenperspectief. Raad voor de Volksgezondheid en Zorg (RVZ)* (in Dutch), Zoetermeer.

3 de Wolf, P., Brouwer, W.B.F. and Rutten, F.F.H. (2005) The Dutch pharmaceutical sector. Controlling costs or increasing efficiency? *Journal of Health Care Management and Planning*, **20** (4), 351–374.

4 Leu, R., Rutten, F., Brouwer, W., Matter, P. and Rütschi, C. (2007) *A Tale of Two Systems; the Swiss and Dutch Health Care Systems Compared.* University of Bern, Bern.

5 Brouwer, W.B.F. and Rutten, F.F.H. (2007) Health insurance: an important area in health services research (in Dutch), in *Handboek Gezondheidszorgonderzoek* (eds T. Plochg, R.E. Juttmann, N.S. Klazinga and J.P. Mackenbach), Bohn Stafleu, Van Loghum, Houten, pp. 208–226.

6 VWS (Ministry of Health, Welfare and Sport) (2005) *Health insurance in the*

Netherlands. The New Health Insurance System from 2006, VWS, The Hague.

7 Schut, F.T. and Van de Ven, W.P. (2005) Rationing and competition in the Dutch health-care system. *Health Economics*, **14** (Suppl. 1), S59–S74.

8 Siciliani, L. and Hurst, J. (2005) Tackling excessive waiting times for elective surgery: a comparative analysis of policies in 12 OECD countries. *Health Policy*, **72**, 201–215.

9 Brouwer, W.B.F. and Schut, F.T. (1999) Priority care for employees: a blessing in disguise? *Health Economics*, **8**, 65–73.

10 Schut, F.T., Gress, S. and Wasem, J. (2003) Consumer price sensitivity and social health insurer choice in Germany and The Netherlands. *International Journal of Health Care Finance and Economics*, **3** (2), 117–138.

11 VWS (2001) *Vraag aan bod (A Question of Demand)*, VWS, The Hague.

12 Committee Structure and Finance of Health Care (Dekker Committee) (1987) *Preparing for Change*, DOP, The Hague.

13 Enthoven, A.C. (1978) Consumer-choice health plan; a national-health-insurance proposal based on regulated competition in the private sector. *New England Journal of Medicine*, **298**, 709–720.

14 Schut, F.T. (2003) De zorg is toch geen markt? Laveren tussen marktfalen en overheidsfalen in de gezondheidszorg (Health Care Is Not a Market, Is It? Switching Between Market Failure and Government Failure in the Health Care Sector.) Inaugural Lecture, Erasmus University, Rotterdam.

15 Van de Ven, W.P.M.M. and Schut, F.T. (1994) Should catastrophic risks be included in a regulated competitive health insurance market? *Social Science and Medicine*, **39** (110), 1459–1472.

16 Lamers, L.M., van Vliet, R.C.J.A. and van de Ven, W.P.M.M. (1999) *Farmacie Kosten Groepen: een verdeelkenmerk voor normuit-keringen gebaseerd op medicijngebruik in het verleden (Pharmacy Cost Groups: Adjusting for Past Medicine Use)*, iBMG, Rotterdam.

17 Lamers, L.M., van Vliet, R.C.J.A. and van de Ven, W.P.M.M. (2000) Farmacie Kosten Groepen: een goede voorspeller van toekomstige ziektekosten gebaseerd op medicijngebruik in het verleden. (Pharmacy Cost Groups: a good adjuster using past medicine use). *Tijdschrift Sociale Gezondheidszorg*, **78** (8), 479–487.

18 Lamers, L.M. (1997) De voorspelkracht van het Diagnose Kosten Groep model. Een evaluatie gebruikmakend van een gezondheidsenquête. (The predictive power of the Diagnosis Cost Group Model. An evaluation using a health questionnaire). *Tijdschrift Sociale Gezondheidszorg*, **75**, 252–262.

19 Van Vliet, R.C.J.A., van Barneveld, E.M., van de Ven, W.P.M.M. and Diagnose Kosten Groepen (1994) Een veelbelovend verdeelcriterium voor normuitkeringen. (Diagnosis Cost Groups: a promising element in risk adjustment). *Tijdschrift Sociale Gezondheidszorg*, **72**, 38–46.

20 Prinsze, F.J., van de Ven, W.P.M.M., de Bruijn, D. and Schut, F.T. (2005) *Verbetering risicoverevening in de zorgverzekering. Van groot belang voor chronisch zieken. (Improvement of Risk-Adjustment Essential for Chronically Ill*, Instituut Beleid en Management Gezondheidszorg Erasmus Universiteit Rotterdam.

21 Enthoven, A.C. (2006) Consumer Choice of Health Plan: Connecting Insurers and Providers in Systems. Paper Presented at Dutch–Flemish Health. Economic Association (VGE) Conference "Consumer Choice: the Right Choice?" VGE, Netherlands.

22 Stolk, E.A. and Rutten, F.F.H. (2005) The "Health Benefit Basket" in The Netherlands. *European Journal of Health Economics*, **6** (Suppl. 1), 53–57.

23 Nederlandse Zorgautoriteit (NZA) (2007) *On the Way to Free Pricing; Yardstick Competition for Medical Specialist Care* (in Dutch), Utrecht.

24 Newhouse, J.P. (1994) Frontier estimation. How useful a tool for health economics? *Journal of Health Economics*, **13**, 317–322.

25 Siciliani, L. and Hurst, J. (2003) Explaining Waiting Times Variations for Elective Surgery across OECD Countries. OECD Health Working Papers No. 7.

26 Swinkels, H. (1996) Waiting for medical treatment: waiting lists (in Dutch). Maandbericht Gezondheidstatistiek, CBS, October.

27 Lindsay, C.M. (1980) *National Health Issues: the British Experience*, Roche Laboratories, Nutley.

28 Cullis, J.G., Jones, P.R. and Propper, C. (2000) Waiting lists, in *Handbook of Health Economics* (eds A.J. Culyer and J.P. Newhouse), Elsevier Science, Amsterdam, pp. 1201–1249.

29 Smethurst, D.P. and Williams, H.C. (2001) Are waiting lists self-regulating? *Nature*, 410, 652–653.

30 Martin, S. and Smith, P.C., (1999) Rationing by waiting lists: an empirical investigation. *Journal of Public Economics*, 71 (1), 141–164.

31 Rutten, F.F.H. (2004) The impact of healthcare reform in The Netherlands. *Pharmacoeconomics*, 22 (Suppl. 2), 65–71.

32 Stolk, E.A., van Buschbach, J.J.V., Caffa, M., Meuleman, E.J.H. and Rutten, F.F.H. (2000) Cost-utility analysis of sildenafil compared with papaverine-phentolamine injections. *British Medical Journal*, 320, 1165–1168.

33 Boston Consulting Group (2000) *Geneesmiddel verzekerd (Medicine Insured)*. BCG, The Hague.

34 Committee De Vries (1999) *Een helder recept (A Clear Recipe)*, Ministerie van VWS, The Hague.

35 Wolters, I. and van Dijk, L. (2000) *Evaluatie Invoering Elektronisch Voorschrijf Systeem (Evaluation Electronic Prescription System)*, Nivel, Utrecht.

36 Van de Ven, W.P.M.M. (1999) De rol van ziektekostenverzekering (The role of health insurance, in *Leerboek Algemene Economie Van de Gezondheidszorg* (eds R. Lapre, F. Rutten and F. Schut), Elsevier/De Tijdstroom, Maarssen, pp. 87–116.

37 Rutten, F.F.H. (1978) *The Use of Health Care Facilities in the Netherlands; an Econometric Analysis*, Rijksuniversiteit Leiden, Proefschrift.

38 Van Vliet, R.C.J.A. (2001) Effects of price and deductibles on medical care demand estimated from survey data. *Applied Economics*, 33 (12), 1515–1524.

39 Holland, J., van Exel, N.J.A., Schut, F.T. and Brouwer, W.B.F. (2007) No-claim: Some pain, No gain. Experiences with the no-claim rebate in the Dutch Health Insurance Scheme (paper), (in press).

40 CPB (2004) Het effect van de invoering van eigen betalingen in de ZFW (The effect of own payments in social health insurance). CPB report, The Hague.

41 Schut, F.T. (2004) Bezuinigingen en hervormingen in de zorg (Cost reductions and reforms in health care) *Economisch-Statistische Berichten*, 89 (4443), 471–473.

42 Kahneman, D. and Tversky, A. (1979) Prospect theory: an analysis of decisions under risk. *Econometrica*, 47, 313–327.

43 Baarsma, B., de Groot, I. and Kok, L. (2004) No-claim prikkelt zorggebruiker minder dan eigen risico (No-claim affects health consumption less than a deductible). *Economenblad*, 27 (5/6), 10–11.

44 Committee on Choices in Health Care (1991) Ministry of Welfare, Health and Cultural Affairs (Report), Rijswijk, Netherlands.

45 VWS (2005) *Health Insurance in the Netherlands. The New Health Insurance System from 2006*, VWS, Den Haag.

46 Rutten, F.F.H. and Brouwer, W.B.F. (2002) Meer zorg bij beperkt budget; een pleidooi voor een betere inzet van het doelmatigheidscriterium (More care with a limited budget; a plea for better use of the efficiency criterion). *Nederlands Tijdschrift Voor Geneeskunde*, 146 (47), 2254–2258.

47 RVZ (2006) *Zinnige en Duurzame Zorg (Sensible and Sustainable Care)*, Raad voor de Volksgezondheid en Zorg (RVZ), Zoetermeer.

48 Stolk, E.A., Poley, M., Brouwer, W.B.F. and van Busschbach, J.J. (2002) Proeftoetsing

iMTA model in *Vervolgonderzoek Breedte Geneesmiddelenpakket* (ed. W.G.M. Toenders), College voor Zorg-Verzekeringen, Amstelveen, Appendix 2.

49 Stolk, E.A., van Donselaar, G., Brouwer, W.B.F. and van Busschbach J.J. (2004) Reconciliation of economic concerns and health policy: illustration of an equity adjustment procedure using proportional shortfall. *Pharmacoeconomics*, 22 (17), 1097–1107.

50 Bellanger, M., Cherilova, V. and Paris, V. (2005) The "health benefit basket" in France. *European Journal of Health Economics*, 9 (Suppl. 1), S24–9.

51 Dickson, M. and Redwood, H. (1998) Pharmaceutical Reference Prices; how do they work in practice? *PharmacoEconomics*, 14, 471–9.

52 Goudriaan, R., Bartelings, H., Thio, V. and Snijders, R.W.D.J. (2006) *Evaluatie van de No-Claimteruggaveregeling (Evaluation of the No Claim Rebate). Kernrapport van de eerste fase*, APE, Den Haag.

53 NICE (2005) *Social Value Judgements. Principles for the Development of NICE Guidance*, NICE, London.

54 Planas-Miret, I., Tur-Prats, A. and Puig-Junoy, J. (2005) Spanish health benefits for services of curative care. *European Journal of Health Economics*, 6 (Suppl. 1), S66–72.

55 Schreyögg, J., Stargardt, T., Velasco-Garrido, M. and Busse, R. (2005) Defining the "health benefit basket" in nine European countries. *European Journal Health Economics*, 6 (Suppl. 1), S2–10.

56 Torbica, A. and Fattore, G. (2005) The "essential levels of care" in Italy: when being explicit serves the devolution of powers. *European Journal of Health Economics*, 6 (Suppl. 1), S46–52.

57 VanVliet, R.C.J.A. (2004) Deductibles and health care expenditures. Empirical estimates of price sensitivity based on administrative data. *International Journal of Health Care Finance and Economics*, 4, 283–305.

58 Williams, A. and Cookson, R. (2000) Equity in health, in *Handbook of Health Economics* (eds A.J., Culyer and J.P., Newhouse), Elsevier Science, Amsterdam, pp. 1863–910.

7
Japan

Akinori Hisashige

Population (million)	127.8
GDP per capita (US$ PPP)	33 500
Health spending as % of GDP	8
Public health spending as % of total spending	81.7
Health spending per capita (US$ PPP)	2 358
Acute care beds per 1000 population	8.2
Practicing physicians per 1000 population	2
Life expectancy at birth (years)	82.1
Infant mortality per 1000 live births	2.8

Strategies used

1. Budget setting (ceilings, fixed payments)
2. Budget shifting (copayments, public budget shifting)
3. Direct/indirect controls (fee schedule and drug price changes, hospital bed and manpower controls)

7.1
Health Care Expenditure Trends in Japan

In 2004, when OECD countries spent an average of 9.0% of their GDP on health care, Japan spent 1% less than the average [1–4]. Japan also ranks below the OECD average in terms of health spending per capita and its growth rate.

7.1.1
Rapid Economic Growth Period

These trends indicate that health care expenditures have been successfully controlled in Japan while main health indicators have been maintained [4–6]. There were two

Cost Containment and Efficiency in National Health Systems: A Global Comparison
Edited by John Rapoport, Philip Jacobs, and Egon Jonsson
Copyright © 2009 WILEY-VCH Verlag GmbH & Co. KGaA, Weinheim
ISBN: 978-3-527-32110-0

Table 7.1 Health care expenditure as a percentage of GDP and average annual growth rate of health care expenditure and GDP (nominal)

Item	1955–1960 High growth	1960s	1970s Stable growth	1980s	1990s Stagnation	2001–2005
Health care expenditure as % of GDP	3.3	3.9	5.1	6.2	7.2	8.8
Annual growth rate Health care expenditure	11.4	19.9	17.3	5.6	3.9	1.9
GDP	14.3	16.3	12.6	6.0	1.3	−0.04
Correlation between healthcare expenditure and GDP per capita	1955–1972 0.998		1973–1990 0.993		1991–2005 0.546	

Source: The Ministry of Health, Labor and Welfare in Japan.

major turning points in the trends of growth in health care expenditures, at about 1978 and 1997, respectively. During the 1950s and 1960s, health care expenditures in Japan were rapidly rising, in fact faster than economic growth (Table 7.1) [6–8]. Although rapid economic growth made it possible rapidly to increase health care expenditures, as well as to expand other social security programs [9], the financial deficit in health care insurance caused much concern among policy makers at the time [10]. The Japanese government responded with a series of measures to control these cost pressures, including several direct financial measures, namely the revision of fee schedules, an increase of the salary limit for payments of insurance premiums, an increase in the premium rate, and copayments among patients.

7.1.2
Stable Economic Growth Period

In Japan during the 1970s, a series of economic recessions caused a serious financial crisis for social security schemes, including health insurance. It had long been recognized that health care expenditures could not go on rising continuously as a proportion of GDP, and policy discussions about health care reform had been started to seek a possible sustainable health care system. Nonetheless, while the growth rate of health care expenditures remained at a high level the rate showed a slight diminishment (see Table 7.1).

From the mid 1970s through the 1980s, the Japanese Government implemented comprehensive and thorough cost-containment measures, including a budget ceiling, a revision of the fee schedule, reductions in drug prices, bundling payments, budget shifting and cross-subsidization. By these means, health care expenditures were substantially contained such that the growth rate was reduced to half of that during the 1970s (see Table 7.1). Average health care expenditures as a percentage of GDP were maintained at around 6%. Although in many different countries a common approach in health care policy has been to combine cost-containment strategies with long-term structural changes to improve value-for-money in health care [2, 11], the Japanese health policy focused mainly on the cost-containment strategies.

7.1.3
Economic Stagnation Period

Health care expenditures have outpaced economic growth over the past decade, starting before the economic downturn of 2001 (Table 7.1). In Japan during the 1990s, the gap between health expenditure growth and the economic growth rate was approximately 2.5%, which was much higher than the average (1%) of most developed countries. Yet, since 2001, this gap has increased [6–8]. Following the collapse of the 'bubble economy' during the early 2000s, the Japanese economy underwent a sudden stagnation which subsequently continued [12] and is the main reason for the current gap between health care expenditures and GDP.

The reduced level of GDP reflects the economic restructuring that is under way, and helps to explain the sharp fall in national revenue available for the health care sector. In responding to this, extensive and intensive cost-containment measures in health care have been implemented, such that the growth of health care expenditure became negative in 2000 and 2002 for the first time in Japan's history.

Although in Japan, since the early 1990s urgent challenges for cost containment have been focused upon and successfully achieved, the efficiency and quality of health care has remained unresolved. Since 2001, organizational and structural reforms for social security schemes, including health care, were proposed as a goal of the Government; in addition, privatization or market-orientated deregulation (with a limited Government role) has also been proposed, but not implemented in the Japanese sociocultural context.

7.2
Health Care System in Japan

When examining health care cost containment in Japan, it is essential to make clear the characteristics of the health care system in the country [4, 13, 14].

7.2.1
Health and Health Care Indicators

The population in Japan (128 million) is the second largest among OECD countries, but has been declining since 2006 for the first time in history. The proportion of elderly is the highest among OECD countries. In terms of health status, Japan has the highest life expectancy among OECD countries, with 82 years for the population as a whole. The infant mortality rate in Japan is the lowest, at almost half the OECD average. Mortality rates for ischemic heart diseases and breast cancer, as well as the prevalence of obesity, are extremely low relative to other OECD countries. On a less-positive note, smoking rates in Japan remain high, but mortality due to lung cancer is very low. The life-style and economic conditions of the population are suggested as the main factors of health improvement in Japan [15].

Whilst the GDP per capita in Japan is slightly higher than the OECD average, the per-capita health care expenditure is slightly lower. In Japan, 82% of health care expenditure is funded by public sources, well above the OECD average. The accessibility to health care in Japan is high, and the number of physician visits per capita in Japan is higher than in many other countries. One characteristic of Japanese health care is that of low staffing [16–18]: the number of practicing physicians per 1000 population in Japan is well below the OECD average, but the number of nurses per 1000 population is slightly higher. However, the ratio of nurses per bed is extremely low, as Japan had the highest number of acute hospital beds among all OECD countries, with more than twice the OECD average. The number of magnetic resonance imaging (MRI) and computed tomography (CT) units per million population in Japan is extraordinarily high. In contrast, however, invasive therapeutic technology has not been widely diffused, compared with that in other developed countries.

7.2.2
Characteristics of the Health Care System

The health care system in Japan is a mixture of centralized governance, social insurance financing and private health care supply. Health care in Japan is highly controlled and regulated by the Ministry of Health, Labor and Welfare (MHLW) [14, 17–21]. The MHLW activities include: health policy development; health information management; monitoring of health status and performance of health sectors; and the regulation of social insurance funds [14, 19]. The flow chart of health care expenditures is shown in Table 7.2. In 2004, more than 80% of revenue was obtained from public funds, and 51% of resources were allocated to hospitals. Half of the expenditures were for medical personnel.

Table 7.2 Flows of health care expenditure (2004); total health care expenditure = $251 billion.

	Hospitals	Clinics	Dental clinics	Pharmacies	Others	Total
Allocation (%)	51.3	24.6	7.9	13.1	3.1	100.0

	Personnel	Drugs	Medical consumptions	Expense	Others	Total
Cost (%)	50.7	15.2	9.6	10.1	13.4	100.0

The Japanese health insurance system has consistently expanded its coverage and benefits, and has universally covered all citizens since 1961 [14, 16–18]. However, there is neither the consumer's choice of health insurers nor the insurer's choice of service providers. At present, the health insurance system is complex and fragmented, consisting of three main schemes with diverse variations among them: (i) the employees' health insurance; (ii) the self-employees' and pensioners' health insurance; and (iii) the Geriatric Health Act [19]:

- The first scheme, for employees, is classified into two subcategories. The first program includes plans for public sector and large company employees, and those managed by the Mutual Aid Association. These cover 32% of the total population; the premiums are 6–9% of salaries, of which the employer pays more than half. The second subprogram is a single government-managed plan for those in small to medium enterprises, which covers 28% of the population. Here, the premiums are 8.2% of the salaries, of which the employers pay a half. Subsidies are provided through the government and amount to 13% of the funding.

- The second scheme, for the self-employees and pensioners, consists of programs established by municipalities, which cover 41% of the population. The premiums are levied on household income, but vary widely by municipality, with a cap per household. Almost 50% of the funding is subsidized by the Government.

- The third scheme consists of plans for persons aged over 75 years. These plans are organized under the Geriatric Health Act, and are run by municipalities. They are financed by a fund of pooled contributions from the other two schemes, and between 20 and 30% of the revenues of the plans are from copayments. With the revenue ceiling in place, the actual copayment rate would be around 14%.

The reimbursement system is rather simple. The nationally uniform fee schedule lists all procedures and products, including drugs, which can be reimbursed by health insurance. No matter how skilled the physician, how prestigious the hospital, or where and by whom the services are provided, the same fee is paid to all providers for the same procedures and products, without consideration of costs. There is no differentiation between physician fee and hospital fee in the schedule, and service providers are paid directly by the insurers. The payments for outpatient care are predominantly on a fee-for-service basis, while inpatient care is paid through a mixture of *per diem* and fee for service. Claims from providers are reviewed retrospectively by a committee of physicians at the local level before reimbursement. The simplicity of reimbursement mechanism, as well as the review system, keeps the administrative costs down.

The fee schedule is revised every two years by the Government, in consultation with the Central Social Insurance Medical Council, which consists of representations from providers, payers and the public. Although this process is extremely complicated, the fee schedule has been a result of political negotiation within the Council, reflecting the relative power balance of the stakeholders. The fee schedule plays a key role in the distribution of health care resources in Japan, as all revenues of health care providers and the Government budget for subsidies are highly dependent on the fee schedule. Although the fee schedule only defines the price of procedures and products – but not their volume – changes in the fees may indirectly influence volume.

As a component of health care expenditures, *drugs* are in an important position, although their proportion to health care expenditure fell from about 40% during the 1980s to less than 20% in 2004. Japanese physicians not only prescribe but also dispense drugs; they usually purchase drugs with 20–30% discounted prices, sometimes with a premium or rebate. There is no *legally imposed* separation between prescribing and dispensing drugs, although separation has been encouraged through an incentive mechanism by the government. Around 50% of all prescriptions are dispensed by physicians.

Japanese physicians who provide primary care are private sole general practitioners (GPs) working in clinics. Specialists mainly work in hospitals and receive a salary [14, 16–19]. The GPs cannot attend to hospitalized patients, while hospital physicians are not allowed to have a private practice as a sideline. The large majority (70%) of hospitals are owned and run by individual physicians but are small in size; however, the remainder are generally large and are public-sector or nongovernment organizations. The latter hospitals receive capital subsidies from the Government and tend to provide most of the advanced care. The existence of for-profit, investor-owned hospitals is prohibited. As provision is weighted towards outpatient treatment, Japan has one of the highest rates of physician visits and lowest of hospital admissions among developed countries. There is little functional differentiation between acute and long-term care among hospitals, and consequently many of the smaller hospitals also serve as long-term care facilities.

7.3
Cost-Containment Strategy in Japan

In Japan, health care expenditure as a percentage of GDP was relatively low at the start of the National Health Insurance system in 1961, but since then has been successfully contained. The problem of health care expenditures has not gone away, however, during the economic stagnation of the past 15 years. Rather, the efficiency and quality of health care have become important health policy issues, and therefore it would be valuable to examine the mechanisms of cost containment, as well as their impacts and limits.

7.3.1
Cost-Containment Strategies

Countries have introduced a number of measures over the past 15 years aimed both directly and indirectly at containing public health care expenditure [22–24]. We adopt a classification of demand and supply [24] based on policies that have been used in Japan, namely budget setting, budget shifting and direct and indirect controls:

- Budget setting
  - Budget ceilings
  - Fixed payment

- Budget shifting
  - Copayment
  - Public budget shifting

- Direct and indirect controls
  - Fee-schedule price revision
  - Pharmaceutical price change
  - Hospital beds control
  - Manpower control

Although a fixed or hard budget has not been explicitly introduced in Japan, budget ceilings set by the Ministry of Finance have crucially influenced the health care budget of the MHLW [25]. Budget shifting is a most common result of reducing public health care expenditure. It is possible that a budget reduction in one area would result in a shift of service demands to another area – to either that of the patients themselves or to that of other parts of the Government's budget. Among the direct controls, fee-schedule price revision and pharmaceutical price changes can be controlled by the Government, whilst for indirect controls, the supply side of health care (which contributes to health care expenditure as a volume) is either regulated or controlled by the Government.

7.3.2
Budget Ceiling

The most important framework for cost containment in Japan is the budget ceiling, whether implicit or explicit. The Ministry of Finance, under the control of the Government, checks and reviews the health care budget plan as requested by the MHLW [25]. The restrictive budget ceiling has been enforced by the Ministry of Finance since the late 1970s. From the starting point of the budget ceiling process, the MHLW has to negotiate its budget plan with the Ministry of Finance, with the budget ceiling being basically set as the maximum limit of budget requirement in each Ministry and not necessarily concerning the contents and allocation of the budget. There are two processes in the fee-schedule revision: (i) a process which decides the total fee-schedule rate revision; and (ii) a process which deals with the allocation of the health care budget and revision of each fee-schedule for all medical practices. Therefore, the budget ceiling set by the Ministry of Finance represents the core of the first process of the fee-schedule revision by the MHLW, and is primarily an important factor for deciding total health care expenditure. The budget ceiling is related to economic conditions in recent years. The GDP or income is indicated as a key determinant of health care expenditures, contributing over 90% to its variance, as documented in many cross-national and longitudinal studies [26–31].

Cost containment by employing a budget ceiling is a dynamic process, where the health care sector responds to changes in GDP through the organizational and financial decisions among stakeholders [28]. The Ministry of Finance and the MHLW quickly respond to changes in recent economic growth, and effectively regulate health care expenditure through the budget ceiling. The relationship between health care expenditure and economic growth became weaker following the economic stagnation period of the 1990s (see Table 7.1), generally because although the health care expenditure was steadily increased the GDP was unusually stagnant. Since 1997, however, the Government has attempted to revise the budget decision process by setting up the Fiscal Structural Reform Committee to more strictly control expenditure on public affairs and social security.

7.3.3
Fee-Schedule Revision

This budget plan is based on the total fee-schedule price rate revision, with the first process – a political negotiation among stakeholders (e.g. the Government, insurers, providers) – usually occurring biannually. The revision of the total fee-schedule price rate is determined by considering both macro-level factors (e.g. economic growth, commodity prices, wages) and micro-level factors (e.g. revenue and expenditure among medical institutions). After setting this revision rate, the fee-schedule for all medical practices is revised. The impact of the fee-schedule revisions is simulated using the results of the Survey of Medical Care Practices in Social Insurance, and checked for correspondence with the budget plan.

While the fee-schedule revision directly changes only the price of medical care, it will indirectly impact on the volume of medical care through economic incentives [18, 32]. These factors should be integrated into the models to estimate the impact on health care costs. Although the market prices for drugs, medical consumption and commission charges for tests are surveyed and considered, most of the other fees are decided on by integrating the requests of stakeholders, the financial conditions among medical institutions, and the policy decisions. The fee-schedule has fully reflected neither the costs nor outcomes of medical care, which have not been evaluated. It would be impossible to evaluate the fee-schedule setting and revision in terms of an efficient allocation of health care resources. However, surpluses or shortages of medical care, from the perspective of health policy makers, have repeatedly been examined in each revision through the above-mentioned survey [32].

The decision rules for revisions are not well formalized, and are rather complex. Formally, this decision has been made by the MHLW with the advisory board, the Central Social Insurance Council. Although this council served as a proposal board for a brief period during the late 1960s, in the late 1970s the sliding-scale approach to prices and wages was adopted for rationalizing decisions and avoiding conflicts among stakeholders. However, as this approach resulted in a rapid increase in health care expenditures the present rule was introduced in 1981. At the same time, in relation to the budget ceiling, the MHLW began to maintain that the growth rate of health care expenditures should be held within the economy's growth rate. Since the end of high economic growth, the Japanese Medical Association has been granted a reduced political power, and the MHLW has taken initiatives with the fee-schedule revision. Moreover, this claim has been treated as a policy goal since 1984, and therefore the strategy for fee-schedule revision has been shifted from one of making costs relevant to one of global cost containment. It should be noted that, since the 1980s, the growth rate of health care expenditure has been much lower

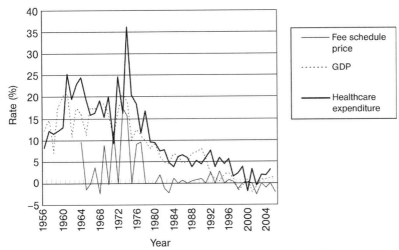

Figure 7.1 Annual changes in GDP, health care expenditure and fee-schedule pricing in Japan.

than that of the Consumer Price Index, or of average wages. This means that although rigorous cost containment was successfully achieved, it could undermine the infrastructure of health care organizations and systems from the long-term standpoint.

As is shown in Figure 7.1, the curves for annual changes of GDP and healthcare expenditure are closely related, with a slight time lag. However, this pattern becomes different during the late 1980s, a bubble economy period, and again after 1991, a period of stagnation. The change of average fee-schedule price is closely related to the trends of GDP and healthcare expenditure, which declined and stabilized at near-zero during the period of stable economic growth, but then became negative after the economic stagnation. The growth rate of fee-schedule revision (an average of two years) is highly correlated to that of healthcare expenditure (with a one-year lag) (1966–2005; $r = 0.713$). The correlation after 1991 becomes low ($r = 0.580$) but, on the other hand, the growth rate of GDP (an average of three years) and fee-schedule revision (average of two years) (1964–2005) is highly correlated ($r = 0.670$). While the correlation during the stable economic growth from the 1970s to the 1980s is high, it has become much higher since the 1990s. Surprisingly, during the period of economic stagnation (1991–2005), the fee-schedule revision was highly correlated to the growth of GDP, with a four-year time lag ($r = 0.924$). These results show clearly that the government responds quickly to economic conditions and adjusts the fee-schedule, as far as it can. Although the fee-schedule revision has led to the successful control of healthcare expenditure, its impact has been decreasing.

7.3.4
Pharmaceutical Price Changes

The overutilization of drugs is one of the key issues in health care in Japan [17]. During the 1960s and 1970s the proportion of drugs in total health care expenditures was extremely high (57% in 1970), but this has since decreased to 22% in 2005. Japanese physicians are permitted both to dispense and prescribe drugs, and to obtain substantial profits from the price differences between the purchase and official drug prices (fee-schedule for drugs). This behavior had been promoted by the Japanese pharmaceutical industry, which developed and marketed marginally innovative drugs [16]. The pricing formula for new drugs in Japan is essentially based on a comparison with the price of similar drugs on the market, and therefore the price was set at a high level for even noninnovative drugs. However, the drug price has been revised every two years to reflect the market (i.e. purchase) price. Health care providers and dispensers purchase drugs at considerably discounted level of the official drug prices (the fee-schedule for drugs). Drug prices have been continuously declining at a rate of 5–10% whereas, since the 1980s, no major increase in the average fee-schedule has been observed, despite drug prices having decreased in continuous fashion. The recently introduced cost-containment initiative seems to depend very heavily on this direct control mechanism.

7.3.5
Fixed Payment

Since the economic slowdown in 1971, cost-containment measures have been rigorously implemented in Japan. However, health care providers still try to maximize their profit by increasing the volume of services to cope with the reduction in fee-schedule. Furthermore, the overutilization of drugs and laboratory tests was accelerated by the Acts when in 1973 the provision of free services for the elderly was instigated. Subsequently, the reimbursement system based on a fee-for-services became one of the most important issues in health care reform, as this payment inherently leads to high costs. Although one solution to the problem would be to introduce fixed payments, this would encourage the providers to eliminate any 'unnecessary' services.

In 1981, the bundling of fees was introduced in order to reduce the costs of laboratory tests. This combines payments for the number and frequency of tests into a flat fee, and was expanded to cover fees for the clinical interpretation of test results. During the 1980s, when medical innovation was led largely by diagnostic testing, this measure brought about a major impact on the contribution of test costs to health care expenditures. Also, in 1993, a fixed payment per diem, which bundles the fees for hospitalization, drugs and laboratory testing, was introduced as an option to pay for chronic inpatient care among the elderly [20]. Although this had a dramatic effect by reducing the costs of both drugs and laboratory tests, the total costs did not change due to an increase in the costs of nurses and other health professionals. Since then, a variety of fixed payments and 'fee bundling' has been introduced in both inpatient and outpatient care systems.

Since 1998, a fee scheme based on diagnostic-related groups (DRGs) has been tested for acute inpatient care, as a pilot study. As with other innovations, the main aim was one of cost containment rather than of quality improvement. The new payment scheme involves 183 DRGs in ten (mainly national) hospitals, but the results, which were evaluated in 2003, showed little promise [33].

In 2003, a new fixed payment system – Diagnosis and Procedure Combinations (DPCs) for acute inpatient care – was introduced [33]. This differs from the DRG/ Prospective Payment System (PPS), in that the fees are *per diem* and decline if the patient's stays in hospital is extended. The DPCs are also a mixture of fixed and fee-for-service payments. The fixed payment is applied to the hospitalization, laboratory tests, drugs and procedures, but excludes operations, radiation therapy, rehabilitation, and so on. Initially, this system was limited to eight university hospitals and two national centers, but has since been expanded to 86 hospitals. The results of a pilot survey showed a decline in the duration of hospital stay, but there was a simultaneous increase in total health care costs as the costs were shifted from inpatient care to outpatient care [33]. It has been suggested that the application of DPCs would be limited because they require considerable investment in hospital information technology. It should be noted that these schemes are at the introductory stage, and do not represent a comprehensive reform of the payment system.

7.3.6
Payments by Households

In Japan, several types of payment are made by households, the most important being the premium rate in health insurance. In recent years, the premium or sharing rate for employees has increased from 2.8% in 1965 to 3.8% in 2002. During the same period the premium rate for employees in small or medium industry has increased from 6.3/2 % to 8.5/2 %.

Although copayment or cost-sharing is indicated to be associated with reduced utilization, it tends to reduce both appropriate and inappropriate care [34]. There are also doubts about the ability of copayment to control total health care expenditures, as providers can increase activity in areas not subject to copayment. Despite these objections, copayment is still very widely used as a policy instrument for cost containment.

In addition, the copayment in elderly health insurance has increased from a nominal fee to ¥300 for consultation and ¥500 for hospitalization per diem (in 1983), and 10% of total fee (in 2002). Besides these major increases, diverse types of copayment for drugs and meals, as well as a decrease in the charge limit of high medical costs, and so on, have been introduced. A review of research on the impact of copayment in Japan showed that a high price elasticity was not observed, except among self-limiting diseases, by increasing the copayments, and that this is also true among the elderly [35].

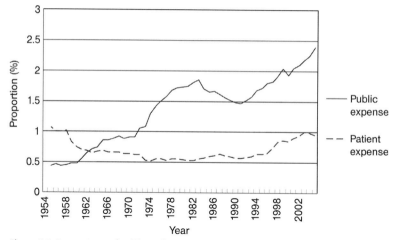

Figure 7.2 Proportions of public and patient expense to GDP.

As shown in Figure 7.2, the proportion of patient expense to GDP decreased slightly after the establishment of national health insurance in 1961, and gradually increased during the 1980s. However, it rapidly increased from 0.5% to 1.0% during the late 1990s, due mainly to the effects of copayment increases. The proportion of public expense has also increased continuously during this period, and represents a serious condition for the Government, which has to propose organizational reforms aiming towards a smaller government. Indeed, the Government's genuine target of cost containment is this public expense, rather than the total health care expenditure. It should be noted that the sudden drop in public expense in 1983 was due to the budget shifting to health insurance for employees, through cross-subsidization by the Elderly Health Act.

Recently, under the financial and policy reforms, the deregulation of restrictions on mixed use and billing of services, insured and noninsured (i.e. balance billing) have been under debate. Balance billing refers to charges made to the patient over and above the reimbursement from health insurance. Under a present rule, a provider cannot be reimbursed for this part, and the total charge must be paid by patients. This practice is allowed only for 12 specific cases, including extra payment for beds with better amenities, and 109 certain high-technology treatments. It has been argued that this deregulation is necessary for providers to meet the increasingly diverse demands by patients. It has also been pointed out that the resulting increase in private payments is the only way to cope with the rising health care expenditure, without increases in contribution rates and taxes. The MHLW and the Japanese Medical Association (JMA) oppose the mixed use of services, and have argued that equal treatment should be provided to all patients, without considering any ability to pay. It would also be politically difficult to offer services according to the ability to pay in an egalitarian society such as Japan.

7.3.7
Public Budget Shifting

The continuity of services from primary to tertiary care and long-term care is poorly coordinated in Japan [17, 32]. The health care system has historically been developed by attaching too much attention to ambulatory care. In line with population aging, a large proportion of hospitals have taken on the function of nursing homes, and this has resulted in providing a considerable amount of unnecessary care or institutionalization of the elderly, as well as an increase in health care expenditure. The fundamental issue for establishing long-term care (LTC) was a lack of resources. A new system of social insurance for LTC started in 2000, with one of the objectives of the system being to transfer LTC hospital beds from health insurance to LTC insurance, together with the expansion of institutional and home care facilities. The result was a budget shift from the MHLW to the public which was estimated to be $1.35 billion (PPP) [36]. In addition, it could serve as a direct measure for reducing unnecessary health care and its associated costs.

7.3.8
Supply-Side Regulation

Both, the number of hospital beds and the length of stay in Japan are extremely high, compared to any other developed country. Figure 7.3 shows the trends in the number of hospital beds. During the 1970s, the number of hospital beds in Japan was slightly higher than in other countries, but in the late 1970s and early 1980s the number was rapidly increased to satisfy a demand for elderly care, as a substitute for LTC. Thereafter, the reduction in hospital bed numbers and the functional differentiation

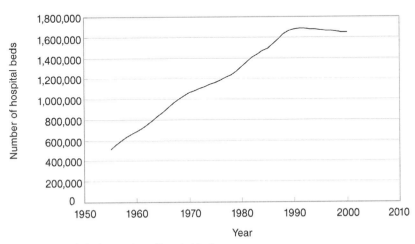

Figure 7.3 Trends in the number of hospital beds.

of hospitals was discussed by the Ministry of Health and Welfare (MHW), and in 1985 an amendment to the Medical Services Law was enacted. This was the first attempt to introduce comprehensive planning to the delivery system in Japan [17, 32]. The main objective was to constrain the increase in hospital beds by setting a maximum number, with the local governments being responsible for regional health planning, including hospital beds. Before this regulation was implemented, many private hospitals initially rushed to open or extend beds to acquire vested rights. However, this increase during a short period was a miscalculation by both the hospitals and the Government, such that the number of beds reached a plateau in about 1991 and later showed a gradual decrease.

Inpatient care has decreased since 1991, together with the introduction of several copayments among the elderly and the decrease in hospital bed numbers. The reduction in beds does not necessarily lead to cost containment, since arranging alternative care requires considerable investment and generates new costs [37, 38]. However, the present number of beds in Japan seems far from the appropriate level, and needs to be adjusted through a reorganization of the health care delivery system, from the perspective of appropriate and efficient care.

With regards to manpower policy, during the 1960s the number of physicians in Japan was not low compared to the OECD countries, although the area variation and shortage of physicians become significant in the late 1960s [39]. In 1970, the Government set the initial target of the number of physicians at 150 per 100 000 population, and since 1973 new medical schools have been aggressively established in each prefecture, such that the number of physicians per population has rapidly increased. Although the target was achieved in 1983, the number of physicians steadily increased beyond that date, and this became a policy concern under health care reform. In 1984, the MHW set up a committee which examined the future supply and demand for physicians and identified four main issues: (i) an over-competitiveness due to a surplus of physicians; (ii) an insufficiency of undergraduate or postgraduate education; (iii) a supplier-induced demand; and (iv) a surplus of physicians in 2000, even by conservative estimate . Consequently, a reduction in the supply of physicians, by 10% minimum by 1995, was recommended, and this was agreed to by the Japanese Medical Association and the Ministry of Education. However, during the 1980s the number of physicians per population in Japan had become low among the OECD countries.

By 1993, the entrance number for medical schools had been reduced by 7.7%, but the goal of 10% was not achieved until later, as private medical schools resisted reducing their student numbers in order to escape financial difficulties. Whilst the number of physicians per population has been increasing, the regional imbalance has not yet been resolved. Furthermore, practice changes due to medical innovation and health transition, with continuing rigorous cost containment, has increased the burden on physicians and created a serious imbalance among specialties and their distribution. In 2007, the Japanese Medical Association admitted for the first time that there was indeed a shortage of physicians. In the past, the Association has mainly represented the perspective of physicians owning private clinics, but it has now begun to consider the critical situation among employed physicians, who comprise the majority. As evidence for the supply and demand of physicians is quite limited, the appropriate level of physicians should be examined from a wider sociocultural perspective [40], including cost containment and cost effectiveness.

7.3.9
Summary

In summarizing cost-containment strategies in Japan, the above-mentioned approaches were successfully implemented during the 1960s and 1980s. There was, however, a hierarchy among these systems. First, a loose and flexible budget ceiling by the Ministry of Finance and the MHLW existed as a budget setting. Second, in achieving this ceiling, fee-schedule and drug prices were extensively revised as direct controls based on negotiations among stakeholders. Most cost-containment successes were realized by these mechanisms. Third, fixed and bundled payments were used in controlling the overutilization of health care. Although various forms of copayment were also introduced for reducing the access and utilization of health care, their effects were limited and tentative for cost containment. With regards to the indirect controls on the supply side, the controls for hospital beds, length of hospital stay and manpower were each unsuccessful. These cost-containment strategies are very similar to those adopted in other developed countries [17].

Efficiency or quality-oriented measures are very limited in Japan. For example, such measures include practice guidelines for doctors, with financial penalties, the establishment of technology assessment institutions, a greater role for technology assessment in coverage and purchasing, activity-related payments, the development and use of sophisticated information systems, investment in developing management competence, and so on [17]. In Japan, although cost containment has been successfully achieved, quality assurance of health care has not been widely used.

7.4
Recent Trends and Future Challenges

The economic stagnation which has occurred since 1991 has led to much greater pressure to contain health care expenditures within the existing framework. Indeed, the tools for regulating and controlling such expenditures have become insufficiently effective, and this has led to much concern regarding any longstanding plans for more effective and radical reforms [11, 41]. Very little discussion has been conducted about the efficiency or effectiveness of health care in Japan, compared to cost containment, even though the quality of health care is a key issue [16, 18]. One reason for this is that the main health care indicators in Japan have been very favorable, while health care expenditures have been low. In the next section, we examine the recent Japanese health policy in detail, with particular attention being paid to cost containment and the prospect for future challenges.

7.4.1
Incremental Reform with Modification

Faced with a deterioration of the health insurance system due to economic stagnation, the need for a comprehensive reform of the health insurance system has been recognized by the Government. The Health Insurance Council, which was set up by the MHWL, began to discuss a future plan in 1993 and, after examining diverse points summarized for discussion, issued the report, "Future Reform of Health Insurance System" in 1995 (Tables 7.3 and 7.4) [42]. The report included several key points for future debate about the basic schemes of health insurance. Traditionally, there are two main strategies in Japanese health insurance: (i) a nationwide social health insurance, and (ii) public funding of health care services . Most items in the report may be classified as budget transfers from health insurance to patients: increases in copayments and premium rates, and decreases of benefits, which are along the lines of traditional measures based on these disciplines.

However, the deregulation of mixed use and the billing of services, to both insured and noninsured, indicated a new direction for introducing and expanding the privatization of health insurance. In the extreme case, there would be the possibility of a transition from a single-tiered system to a multi-tiered system, even though its core was public health insurance. In 1997, the MHW issued the future plan of the health care system, entitled "Healthcare insurance system in the twenty-first century" (see Table 7.3) [43]. Whilst this health care reform plan followed and embodied the former report, fixed payments for hospital care and outpatient care were added. Copayment for physician's fees according to their skill and experience, to a certain extent, and a facility utilization fee based on the care environment, seemed to be applications of mixed use of private and public services. In the same year, after the issue of this plan, the Ruling Parties Council issued "National healthcare in the twenty-first century", which was almost the same as the plan proposed by the MHW [44]. These health care reform plans were created under the comprehensive reform plans including administration, economy, market, finance, social security and education by the Government (the Hashimoto Cabinet) to cope with economic stagnation and large fiscal deficit. Several copayments and sharing rates were implemented later, but mixed-service use was introduced only in limited cases. Therefore, measures for health insurance seem mainly to be still on the traditional course.

7.4.2
Unrealized Scenario for Market-Oriented Health Care

The Economic Strategy Council was established as an advisory board of the Prime Minister in 1998, and issued the final report, "Strategies for Regeneration of the Japanese Economy" the following year (see Tables 7.3 and 7.5) [45]. In the section on health care, this report supported the introduction of market competition to contain health care costs. Market competition included the introduction of free choice of providers for insurers and the Japanese version of managed care. The

Table 7.3 Recent trends of health policy statement, its revision and fee-schedule.

Year	Policy statement	Policy revision	Total fee-schedule revision
1995	• Future Reform of Health Insurance System (Health Insurance Council)	• Increase of copayment of outpatient care in elderly care	–
1996	–	• Increase of copayment of inpatient and outpatient care in the elderly care • Increase of maximum limit of patient expense • Increase of copayment of meals during hospitalization	• +0.8% (+3.4% for medical care, −2.6% for drugs)
1997	• Health care insurance system in the 21st century (MHLW) • National health care in the 21st century (Ruling Parties Council)	• Increase of sharing rate for the insured (20%) • Copayment of drugs for outpatient care • Increase of premium rate for a government managed plan (8.5%) • Increase of copayment for inpatient and outpatient care in elderly care	• +0.38% (+1.7% for medical care, −1.27% for drugs)
1998	–	–	• −1.3% (+1.5% for medical care, −2.8% for drugs)

Table 7.3 (*Continued*)

Year	Policy statement	Policy revision	Total fee-schedule revision
1999	• Strategies for regeneration of Japanese economy (Economic Strategy Council)	• Increase of copayment for inpatient and outpatient care in elderly care	—
2000	—	—	• +0.2% (+1.9% for medical care, −1.7% for drugs)
2001	• Structural reform of the Japanese economy (Council on Economic and Fiscal Policy) • Tentative plan for health care system reform (MHLW) • Issues and perspective for health care system reform (MHLW)	• Copayment for inpatient and outpatient care in elderly care (from fixed charge to fixed rate of 10%) • Changing maximum limit of charge from fixed charge to fixed rate (22% of salary) • Increase of copayment for inpatient and outpatient care in elderly care • Revision of maximum limit of premium rate (9.1%) • Increase of copayment of meals during hospitalization	—
2002	—	• Copayment for the elderly more than 70 years old (fixed rate of 10%) • Revision of subsidiary for medical expenditure for the elderly • Unifying sharing rate among all insurance plan (30%)	• −2.7% (−1.3% for medical care, −1.4% for drugs)

Year			
2003	• Realize a 'revitalized Japanese economy' (Council on Economic and Fiscal Policy) • Basic strategies base on the revision of Health Insurance Act (Cabinet) • Establishment of efficient and transparent health care management as a health care delivery agent (Committee of Future Healthcare Management)	—	—
2004	—	—	• −1.0% (0% for medical care, −1.0% for drugs)
2005	• On the reform of health care system (MHLW) • Tentative plan for structural reform of health care system (MHLW) • Scheme for health care system reform (Ruling Parties Council)	—	—
2006	• Increase of sharing rate for the elderly over 70 years old (30%) • Increase of copayment for meals and stay in chronic care • Increase of maximum limit of charge	—	• −3.16% (−1.36% for medical care, −1.8% for drugs)

Table 7.4 Future Reform of Health Insurance System (Health Insurance Council, in 1995). Symbols: × not realized; △ partially realized; ○ realized (but not necessarily this or next year).

Reform	Outcome
Exclusion of benefits for treatment of self-limiting diseases, meals and bed charge	×
Deregulation of mixed use and billing of services insured and non-insured	△
20% patient sharing rate for the insured	○
10% or 20% patient sharing rate for the elderly	○
Copayment for drugs (30% or 50%)	△
Differentiation of drugs benefit according to their type	△
Reference pricing for drugs	×
Increase of premium rate	○
Utilization of private insurance	×

deregulation of for-profit management of hospitals by corporations, the privatization of public hospitals and free price of drugs, were also recommended. This scenario was the dilemma contrary to the cost-containment strategy, as managed care could not stop the escalation of health care costs [46, 47].

The year 2001 was a turning point. Junichiro Koizumi was elected as Prime Minister, and placed great emphasis on the rule of market, the cutting of public expenditure for social services, the promotion of deregulation and privatization, and shrinking the size of the Government [48]. In 2001, the Cabinet Office and the Council

Table 7.5 Strategies for the regeneration of Japanese economy (Economic Strategy Council, 1999). Symbol: × not realized.

Strategy	Outcome
Introduction of market competition for cost containment	×
Free choice of providers for insurers	×
Introduction of managed care (Japanese version)	×
Deregulation of management of hospitals by a private corporation	×
Privatization of public hospitals	×
Free price of drugs	×

on Economic and Fiscal Policy were established to strengthen the ability for policy development and implementation of the Prime Minister. The Council issued the report, "Structural Reform of the Japanese Economy: Basic Policies for Macroeconomic Management" (Table 7.3) [49]. Although the sections about health care retreated from a perspective of market-oriented health care of the Economic Strategy Council in 1999, the deregulation of management by a private corporation was maintained. The empowerment of insurers was a modification of managed care. Mixed-service use was also revived as a specific item of privatization, and as a fixed payment, while DRG was selected as a new item. It is noteworthy that limiting the growth rate of health care expenditures (in particular for elderly care) within the economic growth rate was picked up, and in the same year the MHLW issued two reports: "Tentative plan for healthcare system reform" and "Issues and Perspective for Healthcare System Reform" [50, 51]. The only new item was the introduction of a control system for the growth rate of elderly care, although the MHLW was in fact against this.

The fate of the scenario for market-oriented health care was as follows. First, the management by a private corporation was limited to a specific structural reform area, for only private services and among hospitals delivering high-technology care [52]. The final report of the advisory committee on the management of health care services set up by the MHLW concluded that management by a private corporation would increase health care costs and that there was no evidence for introducing this change [53–55]. When debates on this issue finally ended in 2003, the strategy of health care reform decided by the Cabinet clearly declared the continuous maintenance of the nationwide health insurance system in the future [55].

Second, mixed-service use was recommended to be extended over the restriction of a specific area for allowing the patients' choice in the strategy decided by the Cabinet in 2003 [56]. However, in the basic strategies several months later in 2003, mixed-service use was again restricted to within high-technology health care, while their rapid acceptance, regardless of specific technologies or hospitals, was discarded [53]. Therefore, at present, there is no great possibility for an unreserved privatization of health care.

Third, DRG/PPS failed in a pilot test in 1998. Instead of this, DPCs – which are a mixture of fixed and fee-for-service – was introduced into major university hospitals in 2003. However, as mentioned above, there is a limited possibility that DPCs will be extended to most hospitals and function as cost-containment tools, as was initially intended.

Fourth, the empowerment of insurers disappeared from the basic strategy in 2003 and the strategy of the Cabinet in 2003 [53, 56]. The action plans initiated by the Regulatory Reform Council did not differ greatly from existing activities, and there were no managerial and human resources for empowerment among insurers. In addition, the reform of the health care system at each community level was not implemented.

Finally, the budget ceiling that had been applied since 1981 was changed by considering only an economic growth rate rather than both GDP and aging [53]. Under economic stagnation, it becomes almost impossible to continuously contain the growth of health care expenditure at a near-zero or minus level. The MHLW pointed out that the main factor for the growth of health care expenditure was the increase in health care for the elderly, and that containing its growth within present economic growth limits would be impossible without sufficient economic growth in the future [22]. The MHLW also suggested that patient expense would become three- to five-fold higher if this containment were to be applied, and that such containment would inevitably result in a deterioration of health status and quality of health care [57]. While these arguments seemed to effectively impact on the discussion, this containment measure has not yet been adopted.

As discussed above, the debates about radical health care reform or the introduction of market-oriented health care were ended in 2003. Despite widespread acknowledgement that radical reform was urgently needed, Prime Minister Koizumi could not realize his envisaged health care reform [58].

7.4.3
Muddling Through

In 2005, the MHLW issued its Tentative Plan for Structural Reform of Healthcare System [59]. This report has mainly focused on the measures for achieving appropriate health care expenditures, rather than simple cost containment as recommended by the Council on Economic and Fiscal Policy. It was a policy extension of the discussion materials mentioned before, and did not include the items for structural health care reform.

For achieving appropriate health care expenditure, as mid- and long-term measures, the prevention of chronic diseases due to lifestyle and a reduction in the length of hospital stays were identified and achievement goals set, despite limited evidence in these areas [60–62]. As short-term measures, various copayments are planned to be introduced, particularly among the elderly. The revision of fee schedule prices in order to reduce short hospital stays, the differentiation of hospital functions, and home care at the end of a patient's life were also considered. These measures are on the lines of traditional cost-containment approaches. While mixed-service use was planned to a large extent, it would bring about much debate after its extent and size had been clarified. In addition, the promotion of generic drug use was introduced as a new item to reduce drug expenditures.

In summarizing the strategies, the report estimated cost reductions through measures proposed, and the Ruling Parties Council issued the Scheme for Healthcare System Reform in 2005, which mostly followed the plans of the MHLW [63]. During the next year, the Act in relation to the Healthcare System Reform based on this scheme was decided by the Cabinet. It is likely that the scheme for health care system reform will shape health care during the next decade.

7.4.4
Future Challenges

It is reasonable that health care expenditures in Japan are increasing more rapidly than economic growth under economic stagnation, as the pressures from population aging and the rapid advance of medical technologies, as well as rising public expectations, have continued. In this situation, cost-containment strategies have become neither desirable nor feasible. The cost containment of health care expenditures has been so successful in Japan that the infrastructure of health care has been undermined for several decades, and this is reflected in the frequent occurrence of medical errors and the shortage and geographical imbalance of physicians (in particular obstetricians and pediatricians) and nurses. Therefore, it is inevitable that a reasonable increase in health care expenditure be allowed, rather than try to impose across-the-board cost containment. Only limited possibilities exist for increasing resources in health care; one is to increase premium rates, and the other is to transfer an increase of consumption tax to health care or social security. As both cases are political matters, the acceptance of these resources will depend on the value judgment among people.

Spending more is not necessarily a problem, particularly if the added benefits exceed the extra costs, although ultimately increasing efficiency may be the only way of reconciling rising demands for health care with public financing constraints [11]. In particular, under the fiscal crisis situation, there are many competitors for resource allocation of the Government budget, such as education, defense, pensions and public works. Therefore, it will be necessary for health care professionals and policy makers to assure the quality and efficiency of health care (i.e. value for money) through health care reform and information disclosure. Since 1997, the MHLW has organized several advisory committees on health technology assessment (HTA) and evidence-based medicine (EBM). These committees have promoted the recognition of EBM among academic areas and people in general, although the Government has not established the formal organization of HTA and EBM. These efforts to incorporate systematic evaluation are invaluably important for health care policy decision making. As has been observed in the recent trends of health care financing, evidence-based policy making is still in its infancy. Almost 20 years have passed since the urgent need for a systematic evaluation of the effectiveness and efficiency of health care have been indicated as a solution to challenges in health care in Japan.

References

1 Huber, M. and Orosz, E. (2003) Health expenditure trends in OECD countries. *Health Care Financing Review*, **25**, 1–22.

2 Mossialos, E. (ed.) (1999) *Health Care Cost Containment in the EU*, Ashgate Publishers Ltd, Aldershot.

3 Saltman, R.B., Figueras, J. and Sakellarides, C. (eds) (1998) *Critical Challenges for Health Care Reform in Europe*, Open University Press, Buckingham.

4 OECD (2007) *OECD Health Data 2007*, OECD, Paris.

5 OECD (2005) *Health at Glance, OECD Indicators 2005*, OECD, Paris.

6 Ministry of Health, Labor & Welfare (2006) *National Health Expenditure 2004 Fiscal Year* (in Japanese), MHLW.

7 Schieber, G. (1990) Health care financing trends: health expenditures in major industrialized countries, 1960–1987. *Health Care Financing Review*, **11**, 159–167.

8 Huber, M. (1999) Health expenditure trends in OECD countries, 1970–1997. *Health Care Financing Review*, **21**, 99–117.

9 Abel-Smith, B. (1984) *Cost Containment in Health Care: a Study of 12 European Countries 1977–1983*, Bedford Square Press, London.

10 Yoshihara, K. and Wada, M. (1999) *History of Health Insurance in Japan* (in Japanese), Toyo Keizai Shinpo, Tokyo.

11 OECD Health Project (2004) *Towards High-Performing Health Systems*, OECD, Paris.

12 OECD (2006) *Economic Survey of Japan 2006*, Policy brief, OECD, Paris.

13 OECD (2007) *OECD Health at a Glance, 2007*, OECD, Paris.

14 Health and Welfare Statistics Association (2007) Trends of national health and welfare. *Journal of Health and Welfare Statistics)*, **54** (9), 1–500 (in Japanese).

15 Marmot, M.G. and Smith, G.D. (1989) Why are the Japanese living longer? *British Medical Journal*, **299**, 1547–1551.

16 Hisashige, A. (1997) Healthcare technology assessment and the challenge to pharmacoeconomics in Japan. *Pharmacoeconomics*, **11**, 319–333.

17 Hisashige, A. (1992). Health care delivery, financing system and aging in Japan. *Journal of the Japan Association of Radiologic Technologists (English issue)*, **39**, 27–52.

18 Hisashige, A. (1993). The Japanese health care system at the crossroads. *Journal of the Japan Association of Radiologic Technologists (English issue)*, **40**, 50–64.

19 Health and Welfare Statistics Association (2006) Trends of national insurance and pension. *Journal of Health and Welfare Statistics*, **54** (9), 1–500 (in Japanese).

20 Ikegami, N. and Cambell, J.C. (1995) Medical care in Japan. *New England Journal of Medicine*, **333**, 1295–1299.

21 Imai, U. (2002) Health care reform in Japan, Economic Department Working Papers no. 321, OECD.

22 Mossialos, E. and Grand, J.L. (eds) (1999) *Health Care and Cost Containment in the European Union*, Ashgate, Vermont.

23 Abel-Smith, B. and Mossialos, E. (1994) Cost containment and health care reform, Occasional Paper in Health Policy no. 2,

London School of Economics & Political Science.

24 Abel-Smith, B. (1984) Cost containment in health care, Occasional Papers on Social Administration no. 73, Bedford Square Press, London.

25 Shiroyama, H., Suzuki, H. and Hosono, S. (eds) (1999) *Policy Development Process in the Ministries of Central Government* (in Japanese), Chuo University Publishers, Tokyo.

26 Newhouse, J.P. (1977) Medical-care expenditure: a cross-national survey. *The Journal of Human Resources*, **12**, 115–125.

27 Culyer, A.J. (1989) Cost containment in Europe. *Health Care Financing Review*, **10** (Ann. Suppl.), 21–39.

28 Getzen, T.E. and Poullier, J.P. (1992) International health spending forecasts: concepts and evaluation. *Social Science and Medicine*, **34**, 1057–1068.

29 Getzen, T.E. (2006) Aggregation and the measurement of health care costs. *Health Services Research*, **41**, 1938–1954.

30 Gerdtham, U.G. and Jonsson, B. (2000) International comparison of health expenditure: theory, data, and econometric analysis, in *Handbook of Health Economics*, Vol. 1A (eds A.J. Culyer and J.P. Newhouse), Elsevier, Amsterdam, Chapter 1, pp. 11–54.

31 Getzen, T.E. (1992) Population aging and the growth of health expenditures. *Journal of Gerontology*, **47**, S98–S104.

32 Hisashige, A. (1994) The introduction and evaluation of MRI in Japan. *International Journal of Technology Assessment in Health Care*, **10**, 392–405.

33 Fushimi, K. (2006) *Inclusive Payment System for Acute Inpatient Care: Its Content and Application*, Financial Review (in Japanese), March, Policy Research Institute, Ministry of Finance Japan, pp. 33–53.

34 Mossialos, E., Dixon, A., Figueras, F. *et al.* (eds) (2002) *Funding Health Care: Options for Europe*, Open University Press, Philadelphia.

35 Ii, M. and Bessho, S. (2006) *Basic Positive Analysis of Healthcare and Policy*, Financial

Review, March (in Japanese), Policy Research Institute, Ministry of Finance, Japan, pp. 117–156.

36 Mori, S. (2001) *Short Overview on Medical Expenditure, 1997.5-2001.5* (in Japanese), Japan Medical Association Research Institute.

37 McKee, M. (2004) *Reducing Hospital Beds: What Are the Lessons to Be Learned?* Policy brief no. 6, European Observatory.

38 Hensher, M., Edwards, N. and Stokes, R. (1999) International trends in the provision and utilization of hospital care. *British Medical Journal*, **319**, 845–848.

39 Ministry of Health, Labor & Welfare (2006) The report of The Committee of Physicians' Need and Supply.

40 Simoens, S. and Hurst, J. (2006) The Supply of Physician Services in OECD Countries, OECD Working Papers no. 21, OECD.

41 Saltman, R.B., Figueras, J. and Sakellarides, C. (eds) (1998) *Critical Challenges for Health Care Reform in Europe*, Open University Press, Buckingham.

42 Health Insurance Council (1995) *Future Reform of Health Insurance System* (in Japanese).

43 Ministry of Health and Welfare (1997) Healthcare insurance system in the 21st century, (in Japanese).

44 Ruling Parties Council (1997) National healthcare in the 21st century (in Japanese).

45 Economic Strategy Council (1999) Strategies for regeneration of Japanese economy (in Japanese).

46 Schwartz, W.B. and Mendelson, N.D. (1992) Why managed care cannot contain hospital costs without rationing? *Health Affairs*, **11**, 100–107.

47 Oberlander, J. (2002) The US health care system: on a road to nowhere? *Canadian Medical Association Journal*, **167**, 163–168.

48 Koellner, P. (2006) The Liberal Democratic Party at 50: sources of dominance and changes in the Koizumi era. *Social Science Japan Journal*, **9**, 243–247.

49 Council on Economic and Fiscal Policy (2001) *Structural Reform of the Japanese Economy: Basic Policies for Macroeconomic Management* (in Japanese), the Cabinet Office, Japan.

50 Ministry of Health, Labor and Welfare (2001) Tentative plan for healthcare system reform (in Japanese).

51 Ministry of Health, Labor and Welfare (2001) Issues and perspective for healthcare system reform (in Japanese).

52 Council on Economic and Fiscal Policy (2003) *Realize a "Revitalized Japanese Economy": Basic Policies for Economic and Fiscal Management and Structural Reform* (in Japanese), Basic policy no. 3, the Cabinet Office, Japan.

53 Committee of Future Healthcare Management (2003) The final report, establishment of efficient and transparent healthcare management as a healthcare delivery agent (in Japanese).

54 Devereaux, P.J., Heels-Ansdell, D., Lacchetti, C. *et al.* (2004) Payments for care at private for-profit and private not-for-profit hospitals: a systematic review and meta-analysis. *Canadian Medical Association Journal*, **170**, 1817–1824.

55 Rosenau, P.V. and Linder, S.H. (2003) Two decades of research comparing for-profit and nonprofit health provider performance in the United States. *Social Science Quarterly*, **84**, 219–241.

56 Cabinet (2003) Basic strategies base on the revision of Health Insurance Act (in Japanese).

57 Ministry of Health, Labor and Welfare (2005) On the reform of healthcare system, in particular related to appropriate healthcare expenditure (in Japanese).

58 Editorial (2005) Leaders: a very Japanese revolution; Japan's election. *Economist*, **376**, 12.

59 Ministry of Health, Labor and Welfare (2005) Tentative plan for structural reform of healthcare system (in Japanese).

60 U.S. Preventive Services Task Force (2005) *Guide to Clinical Preventive Services 2005*,

Agency of Healthcare Research and Quality.

61 McKee, M. and Healy, J. (eds) (2002) *Hospitals in a Changing Europe*, Open University Press, Buckingham.

62 McKee, M. (2004) Reducing hospital beds: what are the lessons to be learned? Policy brief no. 6, European Observatory.

63 Ruling Parties Council (2005) Scheme for healthcare system reform (in Japanese).

8
New Zealand

Toni Ashton

Population (million)	4.1
GDP per capita (US$ PPP)	25 300
Health spending as % of GDP	9
Public health spending as % of total spending	78.1
Health spending per capita (US$ PPP)	2 343
Acute care beds per 1000 population	NA
Practicing physicians per 1000 population	2.2
Life expectancy at birth (years)	79.6
Infant mortality per 1000 live births	5.1

Strategies used

1. Global budgets
2. Quasi-market reforms
3. National public agency to manage pharmaceuticals
4. Priority setting for health spending
5. Waiting list management.

8.1
Introduction

Up until the 1980s, neither efficiency nor cost containment had featured very explicitly as objectives of health policy in New Zealand. The main focus of policy had been on planning and developing a network of publicly funded health services, and on trying to ensure reasonable access to these services by the whole population. As medical technology advanced during the 1970s and 1980s, the public hospital system began to come under considerable strain. Waiting lists expanded, waiting times increased, and some regions had difficulty in providing a reasonable standard of care [1]. However in spite of these pressures, total health expenditure remained fairly static during the 1980s at around 6.5–7% of GDP. Although this was slightly

Cost Containment and Efficiency in National Health Systems: A Global Comparison
Edited by John Rapoport, Philip Jacobs, and Egon Jonsson
Copyright © 2009 WILEY-VCH Verlag GmbH & Co. KGaA, Weinheim
ISBN: 978-3-527-32110-0

lower than the OECD average, it was about the amount that would be expected for a country with New Zealand's level of GDP per capita [2].

From 1984, a reforming Labor government had been active in introducing some fairly radical market-oriented reforms into the wider economy in an effort to improve efficiency and productivity. Capital and labor markets were deregulated, export subsidies and trading barriers were removed, government trading departments were corporatized (if not privatized), and new public management was introduced into the public sector. By the late 1980s, the public health system was one of the few sectors that had escaped the attention of the reformers.

In 1987, a taskforce was appointed by the Government to report on the state of public hospitals and related services. The ensuing report – entitled "Unshackling the Hospitals" – drew attention to the lack of incentives for efficiency in the public hospitals. It also noted the poor integration between primary health services and the secondary sector, and the lack of information about the costs of providing services [3]. The taskforce argued that better management (including improved information systems), clear accountabilities and appropriate monetary incentives could achieve efficiency gains of around 30% within the public hospitals. The report recommended major reorganization of the public health system. Although these recommendations were not acted upon at that time, the report effectively placed the problem of inefficiencies in the health system firmly on the political agenda.

In 1990 an incoming National (i.e. conservative) government announced its intention to reform the public health system. It appointed a ministerial taskforce to recommend changes which would ". . . make funders and providers more efficient and more responsive to consumer preferences" ([4], p. 137). This led, in 1993, to a complete restructuring of the public health system, the central feature of which was the separation of the roles of purchasing and provision of services so as to form a sort of quasi-market in which providers would compete for health funds from purchasers. Although many changes were made to this quasi-market structure over the next seven years [5], enhancing efficiency remained a central objective of health policy. However, in 1999, the Labor Party (which was the senior partner in a newly elected coalition government) made it clear that it was ideologically opposed to the quasi-market structure on the grounds that it was commercially oriented, required competitive tendering for contracts, and lacked democratic input. The government therefore decided to replace the quasi-market arrangements with a ". . . non commercial system, with the focus on the provision of quality services" [6].

A new round of restructuring followed, with the stated key aims this time around being to improve population health, reduce health disparities, and to improve the quality of services. While neither cost containment nor efficiency were explicitly included in the stated objectives, a number of elements of the new system provided strong incentives for the efficient use of resources. In addition, some of the features of the previous system aimed at securing value for money were retained and developed further.

This chapter describes and evaluates some of the main strategies that have been used – directly or indirectly – either for enhancing efficiency or for containing costs in

New Zealand during this period of reform from 1993 to the present. Five policy areas are reviewed:

- the use of global budgets, both nationally and regionally,
- the quasi-market reforms of the 1990s,
- the management of pharmaceuticals,
- the setting of health spending priorities more explicitly,
- the management of waiting lists.

A summary of the strategies introduced within these broad policy areas is given in Table 8.1.

To place these strategies into context, the next section briefly describes the main features of the funding and organization of health services in New Zealand. Each of the five policy areas is then discussed. The chapter concludes with a review of the current policy agenda.

Table 8.1 The main strategies for containing costs and improving efficiency of health care in New Zealand.

Year/time period	Strategy	Objective
1980s	Loose global budgets to Area Health Boards	Contain public expenditure
1989	Public hospitals reviewed	Improve efficiency
1992	Core Services Committee commenced work on setting service priorities	Improve efficiency
1993	Primary care funding incorporated into regional global budgets	Contain public expenditure
	Quasi-market structure established	Improve efficiency
	Pharmac established to manage pharmaceuticals	Contain pharmaceutical expenditure
1995	Principles-based purchasing commences	Improve efficiency
1996	New Zealand Guidelines Group established	Improve clinical effectiveness and efficiency
	Introduction of first Clinical Priority Assessment Criteria for patients on waiting lists	Ensure resources go to those in greatest need
1998	Booking systems in place for nonurgent surgery	Ensure resources go to those in greatest need
2001	Global budgets devolved to District Health Boards	Contain public expenditure (and encourage community responsiveness)
	Capitation payments to Primary Health Organizations	Contain public expenditure (and encourage a focus on population health)
2006	Health technology assessment process introduced	More systematic and consistent introduction of new technologies across DHBs

8.2
Overview of the NZ System

In 2003–2004, New Zealand spent 8.5% of GDP on health services [7]. The health system is predominantly publicly funded, with 70% of total funding coming from general taxation and a further 7.8% from a separate compulsory social insurance scheme for the treatment and prevention of accident-related injuries [7]. Approximately one-third of the population also have supplementary private health insurance which accounts for 5% of total expenditure. The remaining 17% of funding comes from out-of-pocket payments.

Public funding pays for outpatient and inpatient services in public hospitals, mental health services, maternity services and public health services, all of which are provided free of charge to the whole population. In addition, the Government fully or partially subsidizes primary health services, pharmaceuticals, long-term care, disability support (i.e. social care) services, and dental care for children. Dental and vision services are not usually publicly funded for other population groups unless the problem is accident-related. Copayments apply for both general practice consultations and pharmaceuticals. However, between 2002 and 2007, the government increased subsidies for these services quite significantly for the whole population, and so copayments have generally declined in recent years [8]. Private health insurance is purchased primarily to secure faster access to nonurgent hospital services where there are public hospital waiting lists. More comprehensive insurance policies also cover primary care copayments, plus dental and vision services.

Since 2001, the organization of services has been decentralized to 21 district health boards (DHBs) (Figure 8.1). The Ministry of Health allocates tax funds to the DHBs primarily in the form of global budgets determined by a risk-adjusted population-based formula. The DHBs are then responsible for either providing health services directly through their 'provider arm', or purchasing services from nongovernment providers via contracts (called 'service agreements'). DHBs are responsible for planning and funding all personal health services, plus disability support services for people aged 65 years and over for those living in their region. The Ministry of Health has currently retained responsibility for disability support services for younger people and public health services which it purchases directly from providers. However, in time these funds may also be devolved to the 21 districts. The Ministry of Health is also responsible for advising the Minister of Health on National Health strategy, for negotiating strategic and annual plans with the DHBs, and for monitoring their performance against these plans.

The whole system is guided by the New Zealand Health Strategy and the New Zealand Disability Strategy that set out the over-arching direction for health policy [9, 10]. These strategies have also guided the development and implementation of more detailed service, health issue and population-group strategies and action plans. DHBs must comply with these strategies and the national priorities that are set within them when planning their own local services.

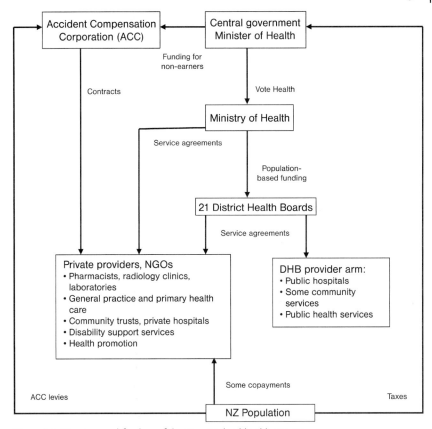

Figure 8.1 Structure and funding of the New Zealand health system.

8.3
Global Budgets

The single most important strategy for containing overall health expenditure in New Zealand has been the setting of a national global budget for health by government ministers as part of the annual budget cycle of the central government. This funding allocation (called 'Vote Health') accounts for around 67% of total health expenditure. The Ministry of Health then reallocates most (almost 80%) of this money to DHBs via the population-based funding formula. This means that just over half of total health expenditure (and about two-thirds of public health expenditure) in New Zealand is effectively subject to a capped budget.

Prior to 1993, the majority of Vote Health was distributed via global budgets to 14 area health boards which were responsible for the provision of public hospital services, public health services and some community-based health services. However, the budget constraint imposed on these area health boards by their global budgets was rather loose, and many incurred deficits that were not accounted for in the National Health accounts. Primary health services and disability support services were funded directly by the Department of Health and Department of Social Welfare, respectively, often on a fee-for-service basis. This meant that any budgetary cap imposed at the national level also had very little impact on the providers of these services. However, from 1993–1994, two important changes occurred. First, the government made public hospital deficits more transparent by including them in the national accounts. Second, funding for all personal health services (including primary health services) and disability support services was pooled into a single funding stream as part of the quasi-market reforms (as discussed below). This meant that these services were effectively drawn under the umbrella of an annual global budget. However, many primary health services (particularly general practice services) continued to be paid on a fee-for-service basis until the early 2000s, when capitated budgets were introduced. Thus, any constraining effect imposed on these services by the national global budget was initially very minimal.

Expenditure under Vote Health has accelerated in recent years, increasing from 15.5% of total government expenditure in 1993–1994 to 20% in 2005–2006 [11]. Much of this increase is due to the transfer of funding and responsibilities for services from other government departments (most particularly the transfer of disability support services from the Department of Social Welfare) [11]. However, there has also been a real increase in government spending on health per capita, especially since 2001. This reflects, in part, a deliberate policy on the part of the Government to increase investment in the public health system. A significant proportion of the additional funds has been earmarked by the Government for the development of primary health services and for increasing subsidies (and hence reducing copayments) for these services [12, 13].

If the transfers from other government departments are excluded, and deficit financing is included, then Vote Health per capita (in NZ$ 2003/2004) has increased in real terms from NZ$ 1511 per person in 1993/1994 to NZ$ 2064 in 2003/2004 [11]. This is an average increase of just over 3% per annum over the 12-year period. However, when expressed as a percentage of GDP, government funding has remained fairly stable (at around 5.7–5.8%) since 1998/1999.

The Ministry of Health now places considerable pressure on the DHBs to work within their global budgets. The DHBs were established under the New Zealand Public Health and Disability Act (2000) which requires them to operate in a 'financially responsible manner' (Section 41). This is defined in terms of maintaining long-term financial viability, covering all annual costs from net annual income, acting as a successful going concern, and prudently managing assets and liabilities. In line with these statutory requirements, the Minister has made it clear to DHBs since their establishment in 2001 that they are expected to cover all annual costs from their net income. Any DHBs incurring deficits are monitored closely by the Ministry of Health, and financial penalties may be incurred (such as payment at the end rather than the beginning of the month). The Ministry also works closely with these DHBs to find ways of reducing their deficits. At the end of the year to June 2004, the combined deficit of all of the DHBs amounted to NZ$ 58 million [14]. However, by the year ending June 2007, financial performance had improved significantly in almost all of the 21 DHBs and, although eight remained in deficit, the 21 DHBs together reported a combined surplus of NZ$ 4 million – some NZ$ 36 million better than projected.

In summary, global budgets for publicly funded services have been important historically in containing total health expenditure. However, they have become increasingly effective in containing costs in recent years as the Government has been more proactive in both encouraging and assisting DHBs to work within their budgets.

8.4
The Quasi-Market Reforms

As noted above, it was not until the 1980s that perceived inefficiencies of the public health system became a focus of health policy in New Zealand. Data deficiencies meant that it was difficult to collate any hard evidence about efficiency levels. Notwithstanding these difficulties, a report prepared by Treasury entitled "Performance of the Health System" found some evidence of improvements in efficiency levels within the area health boards during the late 1980s [15]. This included reductions in the average length of stay, increased surgical throughput, and a decrease in the number of public hospital beds. There were, however, some wide variations in performance across Area Health Boards (AHBs). The Treasury concluded that:

> "On balance we are inclined to view the public secondary health
> sector as having made some improvement in technical efficiency over
> the last three to four years . . .. Considerable variation exists between
> different AHBs. This suggests that some boards could operate more
> efficiently than they are at present." ([15], p. 28 and pp. 38–39).

This rather benign conclusion hardly seems the basis for any major restructuring of the health system aimed at improving efficiency. Nevertheless, an incoming government with a pro-market agenda and an appetite for reform very quickly appointed a ministerial taskforce to: "Identify and investigate options for defining the roles of the Government, the private sector, and individuals in the funding, provision and regulation of health services" ([4], p. 138). The report of the taskforce [4] proposed a completely different organizational structure for the public health system, and this led to one of the most radical and rapid reforms of a public health system ever undertaken. Three key aims of this restructuring were to:

- improve access to a health system that is effective, fair and affordable;
- encourage efficiency, flexibility and innovation in health care delivery; and
- widen the choice of hospitals and health services for 'consumers' ([4], p. 3).

The AHBs were abolished on the same night that the taskforce report was released, and commissioners were appointed to oversee the implementation of a quasi-market structure in which separate organizations became responsible for the purchasing and provision of services. Funding for all personal health and disability services was pooled and, from July 1993, these funds were distributed among four newly established Regional Health Authorities (RHAs) whose role it was to purchase health and disability support services for the population living in their region. The public hospitals and other personal health services previously provided by AHBs were reconfigured into incorporated companies called Crown Health Enterprises (CHEs). Although still publicly owned, these organizations were structured as for-profit enterprises each with two shareholders – the Minister of Crown Health Enterprises and the Minister of Finance. The CHEs were required to compete with private providers to win contracts from RHAs for the provision of personal health and disability services. A parallel structure with a single public purchaser (the Public Health Commission) and multiple (public and private) providers was also initially set up for the purchase and provision of public health services.

The economic theory underlying this quasi-market structure was that competition between providers would secure the desired incentives to stimulate improved technical efficiency within the CHEs and other providers. The taskforce also recommended that, once the basic structure had been established, individuals should be able to take their share of public funding to alternative (public or private) purchasers. This in turn would stimulate purchasers to respond to the needs and preferences of their members, thus encouraging improved allocative efficiency.

The full story of these reforms has been recounted in detail elsewhere [1, 16]. The story is complex and revolves around the difficulties associated with implementing a completely new structure that was strong on theory but weak on both detail and the practicalities of implementation. The election of New Zealand's first coalition government in 1996 (following a replacement of the first-past-the-post electoral system with proportional representation) added another layer of complexity because one of the minority coalition partners which held the balance of power (the New Zealand First Party) was ideologically opposed to a public health system that had a focus on competition and profits. The outcome of all this was that the final structure differed from that originally envisaged by the taskforce. A number of their original proposals were never enacted (including the plan to introduce alternative purchasers), plus many changes were subsequently made to the original structure, especially after the 1996 election. These changes included:

- Abolition of the Public Health Commission and the parallel structure for purchasing and providing public health services (in 1995).
- Abolition of the position of Minister of CHEs and a reincorporation of these responsibilities into the role of Minister of Health (in 1996).
- Replacement of the four Regional Health Authorities with a single national purchaser, the Health Funding Authority (HFA) (in 1997).
- A reconfiguring of CHEs as not-for-profit Crown-owned companies renamed 'Hospital and Health Services' (in 1996).

These, and other, on-going changes to the overall structure and to the organizations within it, together with the fact that numerous other changes were occurring simultaneously within the health sector and the wider economy, exacerbated attempts to undertake any systematic analysis of the impact of the quasi-market structure on efficiency. There does, however, appear to be some consensus among commentators that, while there were some efficiency gains, these were in part a continuation of the gains that had been occurring prior to the restructuring. Moreover, any gains in efficiency were significantly less than expected [1, 17–21].

Many reasons were put forward for this disappointing performance. These included: (i) the absence of competition for many services because public hospitals had geographic monopolies; (ii) the high political costs associated with public hospitals going out of business or closing down services; (iii) the additional transaction costs associated with contracting; (iv) the need for the CHEs to invest some of their budgets in assets and infrastructure that had become run down under the area health boards; and (v) the fact that, in the earlier years at least, it was probably too early to expect efficiency gains because the effects of restructuring would take ". . . 4–8 years to work through" ([18], p. 101) . It was also noted that the original information upon which the expected efficiency gains had been based was poor and the analysis flawed, and so the size of any potential gains had probably been overstated [21].

In summary, the quasi-market structure that was in place in New Zealand from 1993 to 1999 was less effective in achieving efficiency gains than its proponents expected. Although other benefits were reported from the restructuring (e.g. better information about the costs of services, improvements in accountability of providers, and innovations in service development) the experience in New Zealand provides little support for the notion that introducing market-like incentives into a health system will secure gains in efficiency. It is important to note, however, that a system with differences in, for example, institutional arrangements, management structures, geographic distribution of providers, accountability mechanisms, contracting practices and so on, is likely to have rather different outcomes to those reported from New Zealand.

8.5
Management of Pharmaceutical Expenditure

One of the legacies of the quasi-market period has been the Pharmaceutical Management Agency Ltd (Pharmac). Towards the end of the 1980s, the Government became concerned about the growth of state expenditure on community pharmaceuticals which had doubled in nominal terms during the 1980s and increased from 10 to 15% of the total health expenditure [1]. Part of this increase was due to the prices being paid in New Zealand, many of which were known to be higher than those being paid in Australia [22]. The Government set up a Drug Tariff section within the Department of Health to negotiate with the pharmaceutical industry, and a form of reference pricing was adopted.

In 1993, the four RHAs became responsible for the pharmaceutical budget, and formed Pharmac – a jointly owned but independent company which would take over the tasks of the Drug Tariff section and manage the national Pharmaceutical

Schedule on their behalf. The Pharmaceutical Schedule is a type of public formulary which lists those drugs which are subsidized by the Government for community-based prescribing. Pharmac's role is to make decisions about which pharmaceuticals will be listed on the Pharmaceutical Schedule, the subsidy level, and prescribing guidelines and conditions. Doctors and other prescribers may prescribe other drugs which have been approved for use in New Zealand but which are not listed on the Schedule, but the full cost of these drugs must be met by the patient. A key aim of this joint venture was to achieve economies of scale through joint purchasing and hence improve the management of government expenditure on pharmaceuticals.

Pharmac has been one of the few agencies to survive the many changes in health policy that have occurred since it was first established in 1993. Now a Crown entity accountable to the Minister of Health, Pharmac's role has expanded to include purchasing hospital pharmaceuticals on behalf of the DHBs, and demand-side management through the education of consumers and prescribers about the responsible use of pharmaceuticals.

Pharmac uses a range of different strategies for managing the drug budget. These include:

- **Decision criteria**: Pharmac uses a set of criteria when deciding whether or not a drug should be listed on the Pharmaceutical Schedule. These include the availability and suitability of existing drugs, clinical effectiveness and risks, cost-effectiveness (using cost–utility analysis), potential benefit for Māori and Pacific peoples, and total budgetary impact [23].

- **Reference pricing**: Drugs in the same therapeutic group are all subsidized at the level of the lowest priced drug in the group. If pharmaceutical companies choose to set the price of their drug higher than the reference price, then the consumer must pay the additional cost. This stimulates price competition on the supply side and encourages prescribers to use the less-expensive drugs in order to keep the costs down for their patients. In one example, Pharmac negotiated a 60% reduction in the reference price for an ACE inhibitor which had a low market share. This resulted in an increase in market share for the company concerned from 2% in 1997 to 47% in 2004, and a savings of NZ$ 30 million a year in government expenditure [22].

- **Targeting**: Restrictions are placed on the prescription of a drug in order to target its use to those who are most likely to benefit. For example, the patient may have to meet predefined criteria or the drug may only be prescribed by specialists [24].

- **Expenditure caps**: A limit is placed upon total annual expenditure on a drug through the negotiation of price and volume contracts. The pharmaceutical company concerned must refund the Government for any expenditure in excess of this maximum limit.

- **Tendering**: Tenders are sometimes put out for off-patent products. For example, a 44% price reduction was achieved following a tender for paracetamol in 1997. A further 34% reduction was achieved a year later when the product was re-tendered for a period of three years [24]. In some cases, tenders are awarded for the sole supply of a product where only one product is listed so that the supplier receives 100% market share in return for a reduced price.

- **Cross-product arrangements**: A company agrees to a new reference price for a drug in one therapeutic category and, in return, Pharmac agrees to list on the Pharmaceutical Schedule another drug produced by that same company in another therapeutic grouping. In the example of reference pricing cited above, Pharmac agreed to list a new statin on the Pharmaceutical Schedule in return for the 60% reduction in the reference price for ACE inhibitors [24].

These and other strategies used by Pharmac have been spectacularly successful in controlling government expenditure on community pharmaceuticals. Between 1993 and 2006, this expenditure increased by an average of just over 2% a year, a rate somewhat less than the 2.4% average rate of inflation for the same period. As a percentage of total health expenditure, pharmaceutical expenditure declined from 14.8% in 1995 to 12.4% in 2005 [25]. Pharmac has estimated that, in its absence, drug expenditure would have been almost three times higher in 2006 (at NZ$ 1595 million) than the actual expenditure of NZ$ 563 million (Figure 8.2) [26].

In spite of this tight control on expenditure, during its 13 years of existence Pharmac has listed 1031 new or enhanced drugs and widened access for a further 210, while 1132 have been either restricted or delisted ([26], p. 7). However, critics of Pharmac argue that cost containment has been achieved at the expense of access to

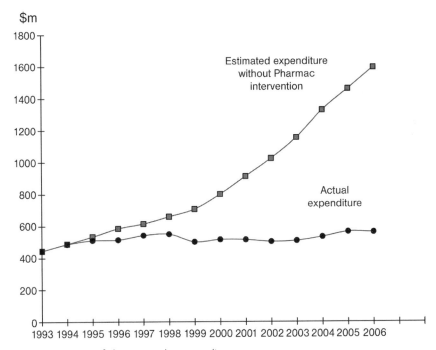

Figure 8.2 Impact of Pharmac on drug expenditure.
(Source: Pharmac (2006) Annual Review 2006, p. 7 [26]).

new and better drugs [27]. Consequently, New Zealand is falling behind other countries in terms of access to new drugs. For example, the Pharmaceutical Industry Taskforce claims that, in the six years to May 2006, Australia subsidized 78 new innovative medicines. While 72 of these medicines were registered in New Zealand, Pharmac agreed to subsidize only 20.

Some clinicians and consumer groups have also called for change in the way that Pharmac controls pharmaceutical spending. In a series of case studies published in the *New Zealand Medical Journal* between May 2005 and June 2006, reported consequences of Pharmac's funding decisions included:

- significant delays in meeting patient needs,
- restrictions on prescribing,
- poor-quality generics and problems associated with sole-supply agreements,
- decisions contrary to international guidelines or evidence,
- denied or reduced access causing increasing costs in other parts of the health system.

In December 2006, the Government announced that it would be developing a Medicines Strategy, and called for submissions from interested parties [28]. While it was emphasized that this was not a review of Pharmac but rather an assessment of the system overall, access to medicines was one of three key objectives of the strategy and Pharmac was the focus of many submissions [29]. It seems likely, therefore, that at least some of Pharmac's activities will come under the microscope during the review process.

In summary, although Pharmac has been extremely successful in containing pharmaceutical expenditure, its role has become increasingly controversial in recent years as some consider that cost containment has been achieved at the expense of access to drugs.

8.6
Priority Setting

The recognition by the Government in the late 1980s that the public health system was not performing as well as it should triggered what has turned out to be an ongoing debate and program of work aimed at determining how health spending priorities should be set. Central to this debate is the question of how efficiency can be improved by channeling limited health resources into those services which provide the greatest benefit and/or which citizens value most highly.

8.6.1
Identifying Core Services

The issue of priority setting first appeared on the policy agenda as part of the proposal to introduce the quasi-market structure [4]. As noted above, the original blueprint for these reforms included a proposal that individuals would have the choice of taking their share of public funds to an alternative purchaser if they wished. Had this proposal been implemented, it would have been necessary for the government to specify some minimum benefit package to which all citizens would be entitled, regardless of their choice of purchaser. Even without any choice of alternative purchasers, the determination of core services would clarify which services the newly established RHAs would be required to purchase on behalf of the government. The identification of a core list of services could also be useful for ensuring maximum possible benefit from any given level of expenditure, for controlling health expenditure, and for ensuring more uniform national access to services [30].

As a first step towards identifying core services, the Government called for public submissions on how a core of health services should be defined [30]. The aim of this exercise was to elicit opinions on whether a detailed priority ranked list of services (as was being developed in Oregon at that time[1]) or some type of more general list was the preferred option. Most submissions indicated a preference for some type of general list. However, a range of other issues emerged which included the importance of widespread public consultation, concerns about the ethical dilemmas associated with limiting or denying access to some services, the possible negative impact on flexibility and innovation, the implied constraint on clinical freedom, and the fact that the process of defining a core would divert scarce health funds away from health services [32].

The Government then established the National Advisory Committee on Core Health Services (or The Core Services Committee as it soon became known) whose task was to identify that set of services to which ". . . everyone should have access, on affordable terms and without unreasonable waiting times." ([4], p. 75). The Committee proposed four principles as the basis for identifying core health services:

- effectiveness
- efficiency
- equity
- acceptability

The Committee reviewed the international literature, convened a series of meetings with experts, stakeholders and the public, and produced a number of publications. However, within a few months it came to the conclusion that a tightly defined core would ". . . either have to be so broad as to be meaningless, or so rigid as to be inflexible and unfair" [33]. The Committee suggested that a better approach would be to make gradual changes to the core of services that was already implicit in the public health system by the fact that were fully or partially publicly funded. The Core Services Committee therefore turned its attention away from trying to decide which services would be provided towards developing principles and guidelines to assist purchasers and providers in making decisions about resource allocation.

---

1) Oregon developed a list of 709 condition–treatment pairs ranked according to their cost–benefit ratio, and to the social values placed by the public on the different treatment categories [31].

8.6.2
Clinical Guidelines

In 1996, the Committee (now renamed the National Health Committee) established an informal network of people called the New Zealand Guidelines Group who have a common interest in the development and implementation of clinical guidelines. This group works with a network of clinical leaders, opinion leaders and consumers to develop and implement tools which promote an evidence-based culture in the provision of health services. In addition to developing clinical guidelines that are appropriate for the New Zealand environment, their work includes the circulation of information about evidence-based practices internationally, and training in guidelines development [34].

Developing guidelines for clinical practice within a publicly funded health service means that the impact on the overall health budget needs to be taken into account. Efficiency is therefore one of the key principles that guides the work of this group. One indication of its success in this regard is illustrated in a paper published in the *British Medical Journal* in 2006 which modeled the potential effectiveness and efficiency of guidelines from five countries including New Zealand on the use of statins for the prevention of coronary heart disease [35]. It found that, while the Australian and British guidelines were most effective in terms of predicted number of deaths avoided over a period of five years, the New Zealand guideline was the most efficient. It potentially could avoid almost as many deaths as the Australian and British guidelines but required treatment for only 12.9% of the population to achieve this benefit compared with 17.3% of the population from the other two countries.

8.6.3
Use of Principles in Guiding Purchasing Decisions

Although the Core Committee did not fulfill its task as originally intended, its consultation process did elicit some useful information about the values that the public considered important when setting health spending priorities. Two additional principles – safety and risk management – were subsequently added to the original list of four, and this new set of principles then formed the basis of a set of purchasing guidelines for the four RHAs [36].

When the four RHAs were replaced by the HFA in 1997, the HFA commenced another program of work (called the Prioritization Project) around the question of how priorities should be set [37]. It gave support to a principles-based approach, to the notion that prioritization should be explicit and transparent, and to the importance of involving the public in decisions about service priorities. It proposed a further modification of the principles to: effectiveness, cost, equity, Māori health[2] and

2) The addition of Māori health reflected the growing political awareness that the Treaty of Waitangi (signed in 1840 between the British Crown and Māori tribes and giving Māori all the rights and privileges of British subjects) had not been honored. Special efforts therefore needed to be made to encourage Māori participation in providing and using services.

acceptability. The HFA also undertook some useful work on how these principles might be operationalized [38].

The abolition of the quasi-market and establishment of 21 DHBs in 2001 means that responsibility for planning and purchasing services has now been devolved to 21 different organizations. The DHBs have been encouraged to develop their own principles and methods for setting purchasing priorities. Most used the HFA's principles as a starting point for discussion and then added or changed the principles as they saw fit. The Ministry of Health provided support by working with the DHBs to develop further the work that had been started by the HFA on how principles could be operationalized [39]. The DHBs have taken different approaches to setting their purchasing priorities, including their processes for consulting with the public. One developed an Oregon-style priority-ranked list by weighting each of its principles and then scoring any new initiatives against this set of weighted principles [40]. Most have taken a less-structured approach, with some paying little more than lip service to the need to set priorities explicitly.

8.6.4
Health Technology Assessment

While the introduction of new pharmaceuticals are rigorously assessed by Pharmac, nonpharmacological technologies have not traditionally been subjected to any formal assessment process in New Zealand. Their introduction into New Zealand has therefore been largely *ad hoc* and inconsistent across the country. In 2006, the Ministry of Health initiated a process for assessing new technologies and service reconfigurations that have regional or national implications and so require some sort of collaborative decision making at the national or regional level [41]. This new framework sets up processes for:

- horizon scanning for new interventions, service changes and potential disinvestments;
- assisting providers to write 'proposals for change' or full business cases;
- developing and consulting on cases for changes in services;
- assisting with the analytical support and access to evidence required to develop a credible case for change;
- making clear decision-making steps and assessment criteria; and
- an annual decision cycle that enables proposals to be prioritized and funding sources identified.

It is too early to report on the success or failure of this new initiative. Unlike Pharmac, the new national and regional groups do not have separate budgets out of which any new technologies can be funded. Instead, their recommendations go back to the DHBs who in turn must decide whether or not a new technology that has been recommended to them should be funded. While the main purpose of this process is to ensure that individual DHBs are not compromised by the decisions of other DHBs, if successful its effect will be to assess new technologies more systematically and rigorously prior to their introduction. This in turn should lead to more efficient resource allocation decisions.

8.6.5
Conclusion

This potted history of priority setting in New Zealand illustrates that, in spite of the amount of time, effort and resources that have been directed towards addressing the question of how priorities should be set in this country, there are still no agreed principles or processes. Even where there has been agreement on how priorities should be set, there have been numerous barriers to implementation. These include difficulties in determining which services or interventions should be assessed for increasing or decreasing expenditure, inadequate information about how the principles might apply to specific services, and lack of funding and skilled resources for undertaking any analysis.

It is impossible to assess whether all of this activity over the past 15 years has improved allocative or technical efficiency. All iterations of the principles for setting priorities have included a principle for assessing cost-effectiveness or value for money. The debate has therefore certainly increased awareness of the need to direct resources into those services that are most efficient. However, problems of applying any principles to purchasing practices have exacerbated the extent to which explicit priority setting might contribute to improving the efficiency of resource allocation. Even when some principles have been applied to purchasing decisions, it has almost always been to assess only new initiatives or for allocating new funds. The principles have still not generally been applied to existing services or programmes which account for the large majority of health expenditure, or to the need for disinvestment in some services.

8.7
Management of Waiting Lists

The existence of waiting lists for the treatment for nonurgent conditions has been an ongoing problem in New Zealand, as it is in most other publicly funded systems where patients are provided with services free of charge. During the early 1990s, in addition to focusing on which services should be given priority, the Core Services Committee directed its attention towards the question of how waiting lists could be better managed. A report commissioned by the Committee recommended that the lists should be replaced by a booking system in which priority for treatment should be determined by agreed criteria based on both need and ability to benefit [42]. A booking system would effectively abolish waiting lists by aligning demand and supply, because only those patients who could realistically be treated within six months of assessment would be booked for treatment. Patients not booked for treatment would be referred back to their general practitioner (GP) for management and reassessed by a specialist at a later date if their level of need increased. A booking system was also expected to reduce the uncertainty for patients about when they might expect to be treated, improve efficiency by allocating resources to those with the greatest ability to benefit, and create a transparent, consistent and equitable means of allocating health resources [43].

Projects soon got underway to develop clinical priority assessment criteria (CPAC) against which a patient's priority for treatment could be assessed. Initially, the CPAC were developed for selected high-volume procedures (e.g. cataract surgery, hip and knee joint replacements), but similar tools were later developed for other medical and surgical interventions [44]. Some CPAC include only medical factors for determining a patient's need and/or ability to benefit. However, in other cases (e.g. cataract and bypass surgery) social factors are also considered to be important determinants of need. Public consultation is therefore sometimes sought to supplement the opinion of clinicians and hospital managers in the selection of criteria [43]. Most of the CPAC tools assign points to various clinical (and sometimes also social) domains; the points are then summed to give an overall priority score [45]. The original idea was that only those patients with a score above some agreed clinical threshold would be booked for treatment.

Studies indicate that the booking system has not operated as originally intended [43, 46–48]. Reasons for this include:

- Development of the CPAC and the booking systems was poorly managed and not coordinated at a national level [43, 45]. Problems included a top-down approach and a resultant lack of buy-in by clinicians; difficulties in reaching consensus over the components and weightings of the clinical and social domains within each CPAC; a lack of collaboration across regions in the competitive (quasi-market) environment that initially prevailed; and a lack of training for surgeons and others in the use of the tools. The result was that CPAC tools tended to be developed locally rather than nationally, and implementation was inaccurate, incomplete and inconsistent.

- CPAC scores do not always correlate closely with indicators of health status using alternative instruments such as the EQ-5D, the SF12 or condition-related health status measures [46]. This may be because the CPAC tools are inaccurate, because staff performing the assessments sometimes manipulate the system to give a higher score to their own patients [43], or because patients themselves may overstate their symptoms if they have some knowledge of the scoring system [45].

- The actual priority given to patients for treatment often does not match their CPAC score [46, 48]. One study also found almost no correlation between CPAC scores and improvement in health status following surgery (i.e. ability to benefit) [49].

- If the intention is to abolish waiting lists by aligning supply and demand, then the threshold has to be determined by the amount of resources available rather than by clinical need. This means that equity of access is unlikely to be achieved, because different financial thresholds apply in different localities.

It is difficult to assess the impact of the booking system on waiting lists and waiting times, not the least because many people who are still waiting for care no longer appear on any waiting list because their level of need has been assessed as being below the threshold for publicly funded treatment. There have also been a series of initiatives directed towards reducing waiting lists over the period since CPAC were first introduced. These have included special injections of funds earmarked by the Government for reducing waiting lists [50], the development of integrated care packages, and changes in referral patterns and practices [45]. In addition, the Government has introduced six-month maximum waiting times for specialist assessments and for treatment following assessment, with financial penalties for any DHB not reaching these targets. One unintended outcome of this latter policy has been a culling of patients from waiting lists by DHBs just prior to the end of each six-month period, with thousands of patients being sent back to their GP for management [51].

Overall, New Zealand's booking system (which has received considerable attention from other countries) has enjoyed limited success to date, at least as far as improving efficiency is concerned. Studies indicate that priority for treatment does not closely align with either need or ability to benefit. Moreover, while fewer people are now 'languishing on waiting lists', many are instead languishing in the community without even the hope of treatment. On the other hand, for those people who are assessed as above the treatment threshold, the certainty of receiving treatment has improved and, in most cases, treatment is provided within six months [52].

8.8
Current Agenda

The change of government in 1999, from a center-right to a center-left coalition, marked a sharp change of direction for health policy in New Zealand. According to the incoming government, improvements in the overall health status of New Zealanders were being hampered by the commercial focus of the quasi-market structure [9]. A more collaborative approach was called for, which focuses on improving the health of the population. Three key goals for the health system were identified: (i) improvement in the health of the population; (ii) reduced disparities in health outcomes, especially with respect to the health of Mâori and Pacific peoples; and (iii) the highest quality health care within the money available [9]. The New Zealand Health Strategy also identified the goals and principles that should be applied to the sector, and specified a set of 13 population health objectives for guiding the prioritization decisions of both purchasers and providers.

While the New Zealand National Health Strategy does not explicitly identify either cost containment or efficiency as objectives of health policy, this new focus on population health was clearly a response to the need to start addressing a number of chronic health problems that could potentially result in a massive increase in health care costs over the next few decades. A focus on population health suggests an approach that reduces health care costs by placing the fence at the top of the cliff – through disease prevention and health promotion – rather than by focusing on improving the efficiency of treatment services. The Strategy noted that: "Improving the population's health means focusing on those factors that most influence health .... [Therefore] it is important for health policy makers, funders and service providers to develop appropriate intersectoral linkages" ([9], p. 5).

Since the release of the New Zealand National Health Strategy, and the restructuring of the health system around the 21 DHBs, there has been a clear reorientation towards population health and intersectoral collaboration throughout all levels of the health system. To give just a few examples:

- The funding and organization of primary health care has been restructured to improve access to primary health services and encourage a population health focus within this sector [12]. The GPs and other primary health providers are now grouped into networks (called Primary Health Organizations, PHOs) which are funded on a capitation basis. Subsidies for GP visits and pharmaceuticals have been increased so that patient fees for these services have fallen and consultation rates are increasing [8]. In addition to assisting providers to improve the quality of their services, the PHOs are active in developing and supporting health-promotion programs based in the community. The new structure has been an important factor in facilitating the development of these programs. For example, capitation funding has shifted the focus away from treatment towards prevention and away from individual providers towards patient management by teams. Capitation has also encouraged providers to develop or improve their patient registers. This in turn has led to a better understanding of the distribution and prevalence of diseases within a community and has allowed PHOs to develop their prevention strategies accordingly.

- In 2003, the Ministry of Health launched a national campaign (called 'Healthy Eating, Healthy Action') to curb the increase in mortality and morbidity associated with obesity, poor nutrition and lack of physical activity. At the launch, the Minister noted that "... we need to act now or face increasing rates of poor health and spiraling health and disability costs." [53]. Numerous national and local initiatives have subsequently been implemented as part of this campaign. Most involve an intersectoral approach that may include schools, churches, workplaces, the food industry, local government and so on, as well as groups within the health sector [53].

- Many DHBs have developed initiatives to address specific diseases that are prevalent in their area. For example, the Counties Manukau District Health Board, which has a very high prevalence of diabetes (associated with its high Māori and Pacific population) has developed a five-year multipronged program called 'Let's Beat Diabetes'. The DHB has taken a lead in mobilizing many different sectors of the community to develop a set of strategies aimed at all points along the diabetes disease path [55].

These examples provide a flavor of the types of policies and initiatives that have been emerging in response to the objective of improving population health. If successful, these initiatives should reduce the risk of disease and/or defer its onset. In the short term, however, as explained above in Section 8.3, government expenditure in the health sector has been increasing, especially in primary health care. This led the department that is responsible for government finances – the New Zealand Treasury – to question whether the additional expenditure that the Government has been directing towards health care is achieving value for money. In 2005, it commenced a work program to examine health expenditure and health sector productivity, and estimated that productivity in public hospitals had fallen by about 7.7% between 2001 and 2004 [56]. However, a lack of centrally recorded data and the poor quality of input data meant that only a minor portion of hospital activity could be included in the analysis. A lack of data also precluded any analysis at all being undertaken of productivity in primary health care, even though this was where most of the additional funding had been directed. Although no firm conclusions could be drawn from these investigations, it did draw attention to the need to find ways of increasing productivity and to improving health sector performance more generally.

A set of policy initiatives directed towards improving health sector performance is being introduced as a result of these findings. One key initiative (in August 2007) has been the introduction of a set of national health targets in 10 key priority areas. No financial incentives or sanctions have been introduced alongside these targets; instead, a collaborative approach is proposed in which the Ministry of Health works closely with the DHBs to plan ways of achieving these targets. The DHBs in turn will work with PHOs and other community-based providers in developing strategies to assist progress towards the targets. The Ministry will monitor performance and, if progress towards the targets is unacceptably slow, consideration may be given towards other means of supporting performance [57]. This is not the first time that national targets have been used in New Zealand, and their potential to carry perverse incentives and produce some unintended outcomes is acknowledged. Nevertheless, the Ministry is of the view that setting specific targets will ". . . help the health sector to focus resources on these areas, lift performance, contribute to overall health improvement and reduce inequalities" [57].

A second strategy for improving health sector performance focuses on the role of the Ministry of Health. Following a major review in 2006, the Ministry was restructured and its role reoriented towards driving 'harder and faster' in some of the health priority areas. A 'change and development' program is now underway to implement other recommendations of the review [58]. The purpose of these changes is to lift productivity and to strengthen the role of the Ministry in managing and improving health sector performance.

A third strategy for improving performance is to review and reconfigure services in priority areas, starting with well-child services, cardiovascular disease and diabetes. The objective of these reviews is to find ways of better directing resources to the most effective areas and to disinvest in less-effective services or components of service. Any changes must be made within existing resources, thus enhancing efficiency [59].

Although other strategies for improving productivity and health sector performance are likely to emerge over time, no significant shift in the direction or focus of current policy seems likely. Indeed, the public health system in New Zealand is more stable at present than it has been for many years. The next general election is at the end of 2008, and the National Party – which is the major opposition party – has stated that, if elected to power, it intends to follow much the same path in health policy as that being followed by the present government [60]. This includes a focus on prevention and early detection as well as maintenance of the current structure and funding arrangements. In being a center-right party, the National Party places considerable emphasis on the need to ensure value for money. Towards this end they are proposing that DHBs should have a greater freedom to use private hospitals to supplement services provided in public hospitals, and a broader range of services should be provided at the primary care level.

8.9
Conclusions

This chapter has discussed five policy areas that have been pursued in New Zealand over the past 15 years in an effort to produce greater value for money or to contain costs in the public health system. Some of these policies have been more successful than others. Global budgets and the strategies pursued by Pharmac have both made important contributions towards containing government expenditure on health care. On the other hand, few obvious efficiency gains emerged from either the quasi-market structure introduced in 1993, or from subsequent efforts to set priorities more explicitly and transparently.

The main focus of attention now is on developing policies which have the potential to reduce costs in the longer term by preventing or deferring the onset of disease. Health status is determined primarily not by health services but rather by numerous social, environmental and economic factors. The success of this approach in containing health care costs therefore depends crucially on the ability of leaders within the health sector to mobilize the support of other sectors in developing and implementing intersectoral strategies aimed at improving the health of the population.

References

1 Gauld, R. (2001) *Revolving Doors: New Zealand's Health Reforms*, Institute of Policy Studies, Wellington.

2 Ministry of Health (2002) *Health Expenditure Trends in New Zealand 1980–2000*, Ministry of Health, Wellington.

3 Gibbs, A., Fraser, D. *et al.* (1998) *Unshackling the Hospitals: Report of the Hospital and Health Related Services Taskforce*, Hospital and Health Related Services Taskforce, Wellington.

4 Upton, S. (1991) *Your Health and the Public Health*, Ministry of Health, Wellington.

5 Ashton, T., Mays, N. *et al.* (2005) Continuity through change; the rhetoric and reality of health reform in New Zealand. *Social Science and Medicine*, **61**, 253–262.

6 New Zealand Labour Party (1999) *Focus on Patients: Labour on Health*, New Zealand Labour Party, Wellington.

7 Ministry of Health (2007) *Health Expenditure Trends in New Zealand 1992–2004*, Ministry of Health, Wellington.

8 Cumming, J. and Gribben, B. (2007) *Evaluation of the Primary Health Care*

Strategy: Practice Data Analysis 2001–2005, Health Services Research Centre, Wellington.

9 King, A. (2000) *The New Zealand National Health Strategy*, Ministry of Health, Wellington.

10 Minister for Disability Issues (2001) *New Zealand Disability Strategy*, Ministry of Health, Wellington.

11 Ministry of Health (2006) *The Annual Report 2005/06 including the Health and Independence Report*, Ministry of Health, Wellington.

12 King, A. (2001) *The Primary Health Care Strategy*, Ministry of Health, Wellington.

13 King, A. (2004) Cabinet Paper: Primary Health Care Strategy – Achieving Low Cost Access, from http://www.moh.govt.nz/moh.nsf/0/EC272299ECCBF6E1CC256C-4F00028D15$File/CabinetPaperLow-CostAccess.pdf (accessed 24 April 2005).

14 Ministry of Health (Various years) District Health Board Financial Performance Reports, from http://www.moh.govt.nz/moh.nsf/indexmh/dhbfp-reports (accessed 21 September 2007).

15 The Treasury (1990) *Performance of the Health System*. The Treasury, Wellington.

16 Davis, P. and Ashton, T. (2001) *Health and Public Policy in New Zealand*, Oxford University Press, Auckland.

17 Crown Company Monitoring and Advisory Unit (1996) *Crown Health Enterprises: Briefing to the Incoming Minister*, Crown Company Monitoring and Advisory Unit, Wellington.

18 The Treasury (1996) *Briefing to the Incoming Government*, Government Printer, Wellington.

19 Ashton, T. (1999) The health reforms: to market and back? in *Redesigning the Welfare State: Problems, Policies and Prospects* (eds J. Boston, P. Dalziel and S. St John), Oxford University Press, Auckland, pp. 134–153.

20 Cumming, J. and Mays, N. (2002) Reform and counter reform: how sustainable is New Zealand's latest health system restructuring? *Journal of Health Services and Research Policy*, 7 (Suppl. 1), S46–S55.

21 Easton, B. (2002) The New Zealand health reforms of the 1990s in context. *Applied Health Economics and Health Policy*, 1 (2), 107–112.

22 Davis, P. (2004) Tough but fair? The active management of the New Zealand Drug Benefits Scheme by an independent Crown agency. *Australian Health Review*, 28 (2), 171–181.

23 Pharmaceutical Management Agency. (2007) *Pharmaceutical Schedule*, Pharmaceutical Management Agency.

24 Braae, R., McNee, W. *et al.* (1999) Managing pharmaceutical expenditure while increasing access. The Pharmaceutical Management Agency (PHARMAC) experience. *Pharmacoeconomics*, 16 (6), 649–660.

25 OECD (2007) *OECD, in Figures 2007*, OECD, Paris.

26 Pharmaceutical Management Agency (2006) *Pharmac Annual Review*, Pharmaceutical Management Agency, Wellington.

27 New Zealand Pharmaceutical Industry Taskforce (2006) *Submission on the "Towards a New Zealand Medicines Strategy" Consultation Document*, New Zealand Pharmaceutical Industry Taskforce, Wellington.

28 Ministry of Health (2006) *Towards a New Zealand Medicines Strategy: Consultation Document*, Ministry of Health, Wellington.

29 Ministry of Health (2007) Towards a New Zealand Medicines Strategy, from http://www.moh.govt.nz/moh.nsf/indexmh/towards-nz-medicines-strategy-sub-missions-jul07 (accessed 30 September 2007).

30 Minister of Health (1991) *The Core Debate: Stage One: How We Define the Core*, Minister of Health, Wellington.

31 Kitzhaber, J.A. (1993) Prioritising health services in an era of limits: the Oregon experience. *British Medical Journal*, 307 (6900), 373–377.

32 The Bridgeport Group (1992) *The Core Debate. Stage One: How We Define the Core*,

Review of Submissions, The Bridgeport Group, Wellington.

33 National Advisory Committee on Core Health Services (1994) *The Core Debater, Issue no. 1,* Wellington.

34 New Zealand Guidelines Group Website (2007) http://www.nzgg.org.nz/ (Accessed 12 November 2007).

35 Manuel, D.G., Kwong, K., Tanuseputro, P., Lim, J. *et al.* (2006) Effectiveness and efficiency of different guidelines on statin treatment for preventing deaths from coronary heart disease: modelling study. *British Medical Journal,* **332** (7555), 1419.

36 Minister of Health (1995) *Policy Guidelines for Regional Health Authorities 1996/97,* Ministry of Health, Wellington.

37 Health Funding Authority (1998) *How Shall We Prioritise Health and Disability Services? A Discussion Paper.* Health Funding Authority, Wellington.

38 Health Funding Authority (2000) *Overview of the Health Funding Authority's Prioritisation Decision Making Framework,* Health Funding Authority, Wellington.

39 District Health Boards New Zealand and Ministry of Health (2005) *The Best Use of Available Resources Prioritisation Framework,* Ministry of Health, Wellington.

40 Hefford, M. and De Boer, M. (2003) Service planning and prioritisation in a district health board, in *Continuity Amid Chaos* (ed. R. Gauld), University of Otago Press, Dunedin, pp. 49–68.

41 District Health Boards New Zealand inc and Ministry of Health (2006) *Service Planning and New Health Intervention Assessment. Framework for Collaborative Decision-Making,* District Health Boards New Zealand Inc. and Ministry of Health, Wellington.

42 Fraser, G., Alley, P. *et al.* (1993) *Waiting Lists and Waiting Times: Their Nature and Management,* Core Services Committee, Wellington.

43 Gauld, R. and Derrett, S. (2000) Solving the surgical waiting list problem? New Zealand's booking system. *International*

Journal of Health Planning and Management, **15** (4), 259–272.

44 Ministry of Health (2008) Elective Services: Guidelines, from http://www. electiveservices.govt.nz/guidelines.html #priority (accessed 14 May 2008).

45 Roake, J. (2003) Managing waiting lists, in *Continuity Amid Chaos* (ed. R. Gauld), University of Otago Press, Dunedin, pp. 107–121.

46 Derrett, S. and Devlin, N. *et al.* (2003) Prioritizing patients for elective surgery: a prospective study of clinical priority assessment criteria in New Zealand. *International Journal of Technology Assessment in Health Care,* **19** (1), 91–105.

47 Dew, K. and Cumming, J. *et al.* (2005) Explicit rationing of elective services: implementing the New Zealand reforms. *Health Policy,* **74** (1), 1–12.

48 Seddon, M. and Broad, J. *et al.* (2006) Coronary artery bypass graft surgery in New Zealand's Auckland region: a comparison between the clinical priority assessment criteria score and the actual clinical priority assigned. *New Zealand Medical Journal,* **119** (1230), U1881.

49 Derrett, S. and Paul, C. *et al.* (2002) Evaluation of explicit prioritisation for elective surgery: a prospective study. *Journal of Health Services and Research Policy,* **7** (Suppl. 1) S14–S22.

50 Hodgson, P. (2006) Funding for 10,000 extra elective procedures, from http:// www.beehive.govt.nz/ViewDocument. aspx?DocumentID=27266 (accessed 30 September 2007).

51 Kiong, K. (2006) *Most Health Boards Cull Waiting Lists,* New Zealand Herald, Auckland, p. A5.

52 Ministry of Health (2007) Elective services: Patient Flow Indicators, from http://www. electiveservices.govt.nz/indicators.html (accessed 30 September 2007).

53 Ministry of Health (2003) Healthy Eating, Healthy Action; Oranga Kai, Oranga Pumau. A Strategic Framework. from www.moh.govt.nz/

healthyeatinghealthyaction (accessed
14 May 2008).

54 Healthy Eating Healthy Action website
(2007) http://www.moh.govt.nz/
healthyeatinghealthyaction (accessed
28 September 2007).

55 Let's Beat Diabetes website (2007)
www.letsbeatdiabetes.org.nz (accessed
28 September 2007).

56 The Treasury (2007) Health Sector
Information Release, from http://www.
treasury.govt.nz/release/healthsector/
default.asp#t2005-344.

57 Ministry of Health (2009) Health Targets:
Moving Towards Healthier Futures,

from http://www.moh.govt.nz/
healthtargets.

58 Ministry of Health (2000) *Building a
Healthy Future*, Ministry of Health,
Wellington.

59 Government Expenditure and
Administration (Exg) Cabinet Committee
(2006) Health Expenditure Review:
Progress Report, Wellington.

60 Ryall, T. (2007) *Better, Sooner, More
Convenient. Health Discussion Paper by
National Party Spokesman, Hon. Tony Ryall
MP*, New Zealand National Party,
Wellington.

9
Sweden

Bengt Jönsson

Population (million)	9.03
GDP per capita (US$ PPP)	32 700
Health spending as % of GDP	9.1
Public health spending as % of total spending	84.6
Health spending per capita (US$ PPP)	2918
Acute care beds per 1000 population	2.2
Practicing physicians per 1000 population	3.4
Life expectancy at birth (years)	80.6
Infant mortality per 1000 live births	2.4

Strategies used

1. Per-case (DRG) hospital reimbursement
2. Reference pricing and generic substitution of pharmaceuticals
3. Health technology assessment.

9.1
Introduction

The Swedish health care system has long been financed by taxes which account for 85% of the total health care expenditure; hence, the control of total expenditures on health care is, in principle, not a major problem. However, when costs increase faster than the tax base, corrective measures must be undertaken, either to increase tax rates or to reduce spending. Although the County Councils,[1] which are responsible for health care, can levy a proportional income tax to finance the service (at present 10.8% on average), the overall expenditure level is determined in negotiations with central

---

1) Eighteen county councils, two regions (Skåne
   and Västra Götaland) and one municipality
   (Gotland) with a medium population of 270 000
   (range 60 000–1.8 million).

Cost Containment and Efficiency in National Health Systems: A Global Comparison
Edited by John Rapoport, Philip Jacobs, and Egon Jonsson
Copyright © 2009 WILEY-VCH Verlag GmbH & Co. KGaA, Weinheim
ISBN: 978-3-527-32110-0

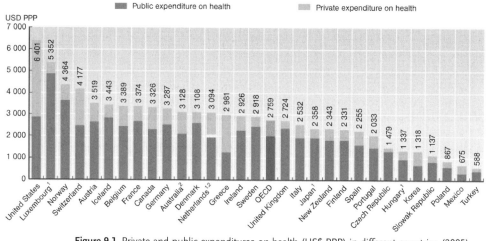

Figure 9.1 Private and public expenditures on health (US$ PPP) in different countries (2005).

government, which also finances about 15% of the costs through general and specific grants to the County Councils. These grants are related to the size of the population, the income level, and indicators of need for services (age, health status, etc.), and aim to equalize access to services in different parts of the country. As the County Councils by law are not allowed to run deficits, and the real opportunities to increase income tax are small in a situation where the tax rate for normal income earners is close to 50%, it is central government – and more specifically the Minister of Finance – that has the ultimate control of the expenditure level. The Swedish County Councils are thus in a similar situation to the public health insurers in Germany and France, where health care is financed by a proportional compulsory fee on the wage bill.[2] In an attempt to not reduce the international competitiveness of the industry, contributions from other taxes play an increasing role in health care financing.

In 1980, Sweden had the highest share (9%) of GDP devoted to health care of all OECD countries, including the US. Today, 25 years later, Sweden has about the same share of GDP for health care, and both the relative and absolute levels of spending and the annual growth rate of health care expenditures between 1995 and 2005 of 3.8% are close to the OECD average. More than half of the OECD countries spend more on health care than Sweden (in PPP adjusted for US$) (see Figure 9.1). The adjustment process has not been easy; during the 1990s the County Councils were constantly in deficit following the downturn in the Swedish economy at the start of the decade. This highlights one of the problems with the Swedish financing system for health care, in that the County Councils have no reserves to meet variations in the tax base, the incomes of the population they serve. Whilst in principle they can borrow money, and their credit worthiness is high, this is often costly when needed and the

2) In fact, private insurers in the US face a similar
   resistance from employers to accept increasing
   premiums.

interest paid must be financed in the annual budget. When the economy takes a downturn, unemployment rises and incomes and taxes are reduced, cost-containment measures must be introduced immediately. Although central government may help to some extent through fiscal policy measures, there are no expectations from individual counties to be 'bailed out' if they do not manage their finances.

In Sweden, most health care institutions, such as hospitals and health care centers, are owned and operated by the County Councils, with the doctors being salaried employees of these institutions. Even the pharmacies are organized in a national corporation, Apoteket AB, such that the pharmacists are salaried employees. However, the link between public finance and public production is being gradually broken. Now, the County Councils are to varying degrees outsourcing the production of both inpatient and ambulatory services to private contractors, both nonprofit and for-profit. Private provision is more common in the three major urban regions, including the biggest cities of Stockholm, Skåne and Västra Götaland. There are also ongoing discussions about the role of Apoteket as a public monopoly for retail pharmacy.

Private insurance play a minor role in Sweden, with less than 2% of the population having a complementary private insurance. However, copayments are used to control moral hazard for all services, including visits to the general practitioner (GP) and to the specialist. Their role as a source of health care financing is limited, however, with the exception of drugs and dental care, where they account for about 20 and 50% of total expenditures, respectively. There has been surprisingly little controversy surrounding the use of copayments, probably because this is a long tradition in Sweden. Moreover, the general social insurance system provides reasonable cover for income losses, and copayments are constructed in such a way as to protect the high users of services. For example, the maximum total annual copayment for pharmaceuticals is €200 per person.

Over a period of about four decades, Sweden has slowly moved from a pluralistic (state, social insurance and county councils) to a single-payer system for health care, mainly driven by the two goals of coordination of services to obtain greater efficiency, and the control of overall costs. This has been achieved while maintaining a decentralized structure of the system, although tensions between centralization and decentralization are built into the system and characterize to a large extent the development of health care policy. Cost control is mainly the responsibility at central level, while efficiency is mainly a responsibility at regional or County Council level. However, several policies at the national level aim at improving efficiency, and cost-containment measures are also common at the County level.

In this chapter we review three policies in Sweden, all of which have goals of both cost containment and improved efficiency. Most policies are characterized by a trade-off between different goals, for example more or better quality services and increased cost; hence, two policies are required to achieve two different goals. The specifics of the chosen policies are that they should – at least in principle – meet two objectives at the same time, namely lower total costs and better efficiency. Improved efficiency may be a result of lower costs for the same effect or better effectiveness for the same cost, but improved cost-effectiveness may also come with increased costs. Actually, improved cost-effectiveness may be a good argument for increased spending, as it indicates better value for money.

The three polices investigated are per-case reimbursement for hospital services based on diagnosis-related group (DRG), reference pricing and generic substitution for pharmaceuticals, and the use of health technology assessment (HTA) to improve decision making on resources in health care. All of these policies have been used over a long period, which makes it possible to evaluate the consequences. All three policies have, potentially, an opportunity to have a desired impact on both total costs and cost-effectiveness. The three policies also focus on the roles of specific Swedish institutions which have been established as a reaction to needs for efficiency and cost containment. Details of the introduction of the policies, together with descriptions of the institutions involved, are listed in Table 9.1.

These three policies have been chosen, among the myriad of 'reforms' that have characterized Swedish health care during the past two decades, because they illustrate the problem of finding policies that balance the need for cost containment and static efficiency, as well as a need for providing incentives for dynamic efficiency – that is, to achieve better and more cost-effective health care through innovation and the introduction of new methods. Hence, attention is focused on policies for 'allocative' efficiency (how much should be spent on health care, and how should the resources be allocated to different services), omitting other policies that are mainly aimed at improving the internal efficiency of the system (doing things right and not wasting resources): improved budget processes and management, reduced waiting times for ambulatory and inpatient care, and so on.

9.2
Paying for Performance: DRG-Based Payments to Hospitals

The Swedish health care system has developed gradually, and many characteristics can only be explained by a long historical perspective. The regional structure for example, which is based on independent County Councils, dates back to legislation from 1862, such that the system is not characterized by revolution. However, the late 1980s and early 1990s was a period that probably is closest to being described as revolutionary in the history of Swedish health care. Suddenly it was possible to have an open and unrestricted discussion about the alternatives to the present system.[3] The reasons for this were several-fold, and economic as well as political, although the most important was a change in attitude to the performance of the system. From being a flagship of Swedish welfare state, with unconditional support from all parts of society, patients, doctors and the general public, questions began to be asked about whether the system performed to expectation. Problems were identified both on the

---

3) See for example Richard Saltman and Casten von Otter (1987), Revitalizing public health care systems: A proposal for public competition in Sweden, *Health Policy, 7,* 21–40 [1].

Table 9.1 Important reforms/changes related to the policies reviewed.

Year	Reform/Change	Aim of the reform/Change
1985	Establishment of Center for Medical Technology assessment (CMT) in Linköping by The County of Östergötland and Linköping university	To undertake HTA studies and research on methods and the role of HTA in decision making
1987	Establishment of SBU, The Swedish Council for Health Technology Assessment	More efficient and cost-effective use of technologies and methods used for prevention and treatment
1991	Introduction of the Stockholm model. The most prominent experiment with models for health care organization based on purchaser–provider split, payment for performance, and consumer choice	Patient choice and improved efficiency; that is, improved productivity and quality while controlling total expenditures
1993	Introduction of reference pricing for off-patent pharmaceuticals, managed by the National Social Insurance Board	Contain public expenditure for pharmaceuticals and improve efficiency; that is, to achieve the same effectiveness at lower cost
1996	National guidelines for different diseases; National Board of Health and Welfare. Based on clinical and health economic documentation	Management through publications of guidelines based on science and established practice to improve efficiency in health care
2001	Establishment of a National Center for Priorities in Health and Social Care (PrioriteringsCentrum); http//e.lio.se/prioriteringscentrum	Undertaking research and information about open priorities in health and social care
2002	LFN, The pharmaceutical Benefits Board, is established to make reimbursement decisions and manage the price regulation scheme, including generic substitution.	Create a rational mechanism for reimbursement of pharmaceuticals to help contain costs and improve cost-effectiveness
2002	Reference pricing abolished and generic substitution introduced.	Contain public expenditure for pharmaceuticals and improve efficiency; that is, to achieve the same effectiveness at lower cost

supply and demand sides, such as low productivity and a lack of responsiveness to demands from patients.[4]

The Swedish situation was of course far from unique: other countries had also entered the same discussions, and there was a growing interest in comparisons with other countries and learning from their experiences, probably in the hope that a 'perfect system' should be somewhere out there.

---

[4] For an in-depth review of the problems and proposed solutions, see Göran Arvidsson and Bengt Jönsson, *Politik och marknad i framtidens sjukvård*, SNS Förlag, 1997 [2]. This book summarizes a research project including many Swedish and international researchers over the period 1990–1997 published in over a dozen books and papers. See also Ref. [3].

The Government initiated a major inquiry in the form of a public committee (The Committee on Funding and Organisation of Health Services and Medical Care; HSU 2000) to prepare for new legislation to change the system. As a starting point for the work of the committee, a group of experts was assigned with the task of defining and evaluating different models for Swedish health care in the future [4]. It was not only at the central level that initiatives were taken; the individual County Councils were also beginning to experiment with new models for health care delivery and financing. One common feature of all these models was a separation of purchaser and provider roles within the County Council. This was a major break from the traditional, integrated model where budgets, with specification of spending on different cost items such as staff and running expenses of different types, were allocated to institutions such as hospitals and health care centers. There was no way to identify exactly for what purposes the resources were used, for example in the treatment of different types of patients. A more detailed description of the reforms is available in Ref. [5].

While all new models were based on the idea of separating the roles of the purchaser and provider, they differed very much in terms of how the relationships between purchasers and providers were structured. One theme which aroused heated debate was to what extent competition between the providers and free choice for consumers/patients to select care givers could be used to improve the performance of the system. It was soon realized that competition and free choice were not easy concepts to put into practice in Swedish health care, with its strong emphasis on equity and consequently public funding. There were also ideological differences linked to these concepts which prevented a national consensus for a reform which included these aspects. Most models therefore had a rather short life span, and none was developed into what could have been a new 'Swedish model'.

The purchaser–provider split remained, however, and with it the focus on how health care providers – and in particular hospitals – should be paid for their services. The development of a Swedish DRG system to classify patients according to the diagnosis and procedures undertaken, as well as other variables that indicated a need for resources (e.g. age and presence of complications) had been started during the 1980s [6]. The main purpose of the DRG system was to describe hospital production or activity in a way that allowed comparisons to be made between hospitals by allowing a standardization for case mix. However, as had occurred in the US, payers for health care were also interested in using this for costing and reimbursement.

Although the system was not fully developed, several County Councils decided to use it as an instrument to improve efficiency and to control costs. The most consequent application was undertaken by the Stockholm County Council; following a decision made during the autumn of 1992, the council introduced DRG-based payments to hospitals as part of a major change in organization and management. This included a purchaser–provider split, a system for internal pricing, and freedom for the patient to choose the hospital where the treatment would be received. This analysis will focus on the use of DRG-based costing, budgeting and reimbursement (payment).

Payment per treated patient, as an alternative to a fixed budget for the hospital, has a number of positive characteristics and incentives. One positive is that it may give the patient freedom to choose a hospital, instead of being directed to a specific hospital. Such a choice has a value if there are a number of hospitals to choose from; this is the case in Stockholm, but not in several other parts of Sweden, where there is only one alternative within a reasonable distance. The reduction in the number of hospitals due to structural changes, and the need for specialization to increase quality and reduce costs, has diminished the choices available. The better information and transport facilities have increased the opportunity for choice, although it does mean that, to a greater extent, these choices concern services offered outside the patient's own County Council, and perhaps even in another country.

Payment per treated patient and free choice provides a hospital with incentives to improve quality; otherwise patients will not choose to be treated there. One problem with this argument is that it is not easy for the patient to observe quality of care, which is necessary in order to make a rational choice. There is also the risk that if the patient cannot judge quality in advance, and if the probability of repeat consumption is low, the hospital can increase the 'profit', or at least their competitive advantage, for a specific DRG by reducing the costs and quality. The experience gained from the Stockholm model was that there was no reduction in quality, but neither was it possible to observe any increase in quality. This indicates that quality of care must be managed by other policies, and that DRG reimbursement in itself is not directly related to quality. A number of other factors are more important, such as internal processes, the training of staff, and management.

DRG-based payment provides an opportunity for the purchaser to directly influence how resources are spent for different types of patients, in a more precise way than when a global budget is used for resource allocation. Instead of increasing the budget for the orthopedic surgery department, it is possible to 'purchase' a defined number of hip or knee replacements. While this is an advantage, similar 'contracts' could be negotiated even without DRG financing. In general, however, DRG payment provides the opportunity to achieve a more transparent view of how the resources are used, while the separation of numbers of units and price/cost per unit provides further information about what drives changes in the total expenditures for the different types of service.

9.2.1
Incentives for Cost Containment

Per-case reimbursement gives incentives to the provider to not waste resources, and for continuous improvements of the processes. Given restrictions on quality, the incentive is to minimize the costs per case through an optimal combination of inputs to the health care production. There is strong evidence that these incentives also work in practice. The introduction of per-case reimbursement in the Stockholm County Council was followed by a dramatic increase in the number of units produced with the available resources. Comparative studies between County Councils that intro-duced per-case reimbursement and those that kept the traditional global budget system also showed that there was an increase in productivity. Estimates indicated a cost reduction per case of 13% through a shift from budget allocation to performance-based reimbursement [7].

The experience reveals that there was a large potential for a more efficient utilization of resources in the Swedish health care system at the time, with about a 20–25% increase in the initial phase. It is more difficult to judge the long-term effects, as the experiment was not continued for long enough to assess such dynamic effects. The reason for this was that the start coincided with a rapid downturn in the Swedish economy, where GDP per capita was reduced each year between 1991 and 1993 for the first time since the Second World War. This forced the County Councils into cost-containment strategies, which included a reduction in payments to the hospital per DRG unit. A certain reduction was not surprising when it was revealed that there was capacity to produce many more units than expected. However, due to the fall in the County Councils' income the reduction was close to the observed increase in volume, which left the providers with no reward for their efforts. The worsening budget situation also created a need for further cost-containment measures, such as restrictions on volumes. Not surprisingly, this eroded any support among the medical staff for the DRG-based per-case payment system.

Whilst it can be concluded that per-case payment is an effective instrument to release the potential for productivity improvements within a system based on budget allocation, the consequence is that volumes increase and, in turn, so does the total cost. This may not have been such a big problem if the economic situation had been favorable, so that additional funds could be transferred to the health care system. However, during a period when it was necessary to cut overall spending, it did not work out very well. While a DRG-based reimbursement system was maintained, the need to control volumes for the individual hospitals meant that the system became a compromise between a per-case reimbursement system and a global budget system. The major problems were how to combine per-case reimbursement with a need to control total costs, and how to combine freedom of choice for the patient with the purchasers' objectives to allocate resources to reach specific objectives, such as changes in the utilization patterns from acute to primary care. It also became clear that, in order to improve quality of care, specific measures were needed that were complementary to per-case reimbursement.

9.2.2
Development After the Mid 1990s

The purchaser–provider split and per-case reimbursement based on the grouping of patients according to DRG was a policy instrument that proved its value and was there to stay. However, it was not a 'quick fix' for improvements in efficiency and quality, and was not particularly helpful in a situation where drastic cost containments were needed.

Productivity decreased during the 1980s, but then improved dramatically during the early 1990s with the introduction of new models for financing and management, as described above. An important change here was also a reform where the responsibility for long-term care patients was transferred from the County Councils to the municipalities, which released hospital capacity to increase the number of patients treated.

However, from the mid 1990s onwards a further downward trend was observed in productivity. Although, resources in terms of the numbers of doctors and nurses have increased, this has not been the case for the number of hospital inpatient admissions and visits to ambulatory care. International comparisons indicate that the Swedish health care system will continue to have productivity problems, mainly resulting from a low utilization of doctors and nurses.

The presently used DRG system, NordDRG, was introduced in 1996 and is similar in all of the Nordic countries. The Swedish version now also includes a classification of ambulatory procedures and psychiatric care. In the nationwide Hospital Discharge Register, all cases in Sweden are grouped in NordDRG. Whilst the hospitals and County Councils mainly use the DRG-based information for management purposes, NordDRG can also function as a prospective payment system, and is used today to reimburse about 50% of Swedish inpatient cases. It is most commonly used for the reimbursement of care from one county to another. When used for internal reimbursement within a County Council, for example in Stockholm, Uppsala, Örebro, Östergötland, Skåne, Halland and Västra Götaland, the reimbursement level is only about 50%, with the remaining costs being covered through a fixed budget.

The DRG classification system is based on the idea that patients in the same DRG group should have about the same resource use. However, it has taken a long time to develop a system for case costing which is complementary to the DRG classification system. The present system in Sweden (cost per patient; KPP) covers about 50% of the hospital inpatient population. Although the system is being continuously improved, it is only used in a minority of counties for budgeting and resource allocation. In those counties who do use it, the role seems to be limited, not systematic, and transparent. The reason for this may be a lack of confidence in the estimates, although a limited active use does not provide any incentives for improvement.

The development of a system which allows payment for performance is of key importance in the development of Swedish health care. It is difficult to see an alternative solution to the chronic problem of low productivity, as improving efficiency and quality through an increased choice of provider for the patients, as well as competition between providers, can only be achieved if an efficient reimbursement system is established. However, such a system must be complemented with additional incentives for quality and, most importantly, it must be managed very closely. Thus, measurement and management are intimately linked, and without timely and correct management information systems for costs and quality, the potential benefits will not be realized.

9.3
Reference Pricing and Generic Substitution

Reference pricing and generic substitution for drugs for which the patent has expired seem to be obvious policy instruments for achieving both greater efficiency and lower costs. Although, drugs that incorporate the same substance may differ in ways that are important for individual patients, overall the effectiveness and side effects can be assumed to be the same. It is thus cost-effective to choose the drug with the lowest price.

Why is this not automatically done in a health care system that is always short of money? Studies in Sweden have shown that unless the patient pays a copayment, doctors do not select the least expensive generic [8]. If a third party, the County Council, pays the difference between a low-priced and a high-priced generic drug, the prescribing physician does not act in a price-conscious way.[5] Consequently, it may be necessary to introduce mechanisms that take this decision away from the doctor and place it somewhere else.

Reference pricing, which was introduced in Sweden in 1993, was constructed so that the reimbursement scheme pays for 110% of the price of the lowest-priced generic. If the doctor prescribes a drug with a higher price, the patient has to pay the difference. The experience in Sweden is that very few patients are prepared to pay this extra cost, as most prescriptions are for patients who are used to receiving free medications. Hence, drugs with a higher price than the generic price will lose most of their market more or less instantly. It is therefore a rational policy to drop the price to the reimbursement level, although for a product with a strong brand name – particularly if the product is used for long-term medication – the seller may consider keeping a high price and a small market share.

When reference pricing was introduced in Sweden in 1993, an initial reduction in the prices of generic drugs covered by the scheme was also observed, together with a total reduction in spending on these products of 430 million SEK. Whilst this was a significant amount of money to have saved, it still represented only about 5% of the total drug spending, which was equal to the annual increase in total drug expenditures at that time. So, the policy had a one-off cost-containment effect, but had no effect on the growth of total health care expenditures. Actually, during the period between 1993 and 2002, when the reference pricing system was replaced by generic substitution, the total increase in pharmaceutical expenditures was more than 10 000 million SEK.

Then why was the reference pricing system abolished? One reason is probably that it was administratively complicated and costly to administer, and indeed this was a stated reason for its abolition in Norway.[6] A second reason was that, after the initial effects, no further savings were achieved. Instead, it was possible to detect a negative impact on the dynamic efficiency in the market [9]. The fact that there were fewer entrants to the generic market after the reference price system had been introduced was a signal that the dynamic competition had been weakened. The reason for this was that new entrants to the market could only gain sales if they priced their product below the reference price. If the price was positioned at or above the reference price,

---

5) An alternative policy would be to include incentives for doctors to take price into account.
Experiments with local drug budgets have been
undertaken in Sweden. Such incentives will
also have effects on choices between drugs
that are therapeutic alternatives, but do not
necessarily contain the same agent. Reference

pricing for such larger classes of drugs ('jumbo
groups'), have been used in other countries
such as Germany, but not in Sweden.

6) Reference pricing was introduced in Norway
in March 2003 for a group of off-patent drugs
with high sales, and abolished at the end of
2004.

then the new entrant lacked any specific argument for doctors/patients to select its product. However, if they priced below the reference price, then a new and lower reference price would be calculated at the next revision and this would force competitors also to reduce their prices. Reference pricing thus acts as a 'compulsory price war', and new entrants to the market must take this into account. As it will be less profitable to enter the generic market, there will therefore be fewer entrants. There are also strong incentives for cooperation between different competing firms in the market, as most will lose from a price war. While reference pricing provides initial price reductions and cost savings, in the long run the benefits are smaller, and may even be turned into losses when compared to a generic market without reference pricing. In such a market we may see different prices for drugs with the same active substance. Although this may seem inefficient at a defined point in time, it can over a prolonged period lead to lower prices and lower total costs.

Generic substitution replaced the reference pricing system in 2002. Generic substitution means that the pharmacy can substitute a prescribed product, being the original brand or a generic equivalent, within the same class, with one at a lower price. There are some qualifications to this practice, the two most important being that the doctor has noted that substitution is not allowed, or the patient agrees to pay the difference between the reimbursement price and the actual price for a higher-cost drug.

One major advantage from a marketing point of view is that new entrants to the market do not need to spend resources to influence doctors to prescribe their brand. They have only to ensure that their product is listed among the substitutes on the market, and that it is priced so that there is incentive for the pharmacy to select it. From an administrative point of view, there is also the advantage that there is no need to control prices for drugs where there are generic equivalents. The only decision needed is the reimbursement level, which will be adjusted according to the price development in the market.

The introduction of generic substitution was seen as a great success, as it was followed by fast and dramatic reductions in the prices of major products for which the patent had recently expired. Price reductions are generally followed by increases in sales volume, but as the effect on volume was rather modest the total sales for the class of drugs also fell significantly. An estimate of the resulting price reductions and cost savings for a number of drugs is shown in Table 9.2. The major reduction in the average price for each class of drugs comes from reductions in the price of the product that was losing its patent. However, statistical analyses show that patented drugs within the same therapeutic area also lost sales.

The cumulative savings from the price decreases and shifts in sales were estimated at SEK 7 billion (ca. US$ 1 billion) during the period 2002–2005; this amounts to about SEK 2 billion annually, or about 10% of the total drug sales in Sweden.

The estimated savings may to some extent have been overstated, however, as there are incentives for those doctors and patients who do not like the prospect of substitution to select a product from a class where there is no generic substitution. As these products are under patent, the price is also generally higher, and the extra costs for this shift in prescriptions must be deducted from the savings. This may to some

Table 9.2 Price changes after patent expiration for four drugs and drug classes

Drug class/ Drug	Date patent expired	Average price per DDD October 2002 (SEK)	Average price per DDD December 2005 (SEK)	Change (%)
Proton-pump inhibitor		13.1	7.8	−41
Omeprazole	March 2003	15.5	5.5	−65
Statin		7.9	2.3	−71
Simvastatin	February 2003	8.5	0.7	−92
SSRI		7.2	2.4	−66
Citalopram	June 2002	7.2 (June 2002)	1.2	−83
Calcium antagonist		4.9	3.1	−35
Felodipine	February 2003	4.4	1.7	−61

DDD, defined daily dosage; SSRI, selective serotonin reuptake inhibitor.
Source: Engström A, Jacob J and Lundin, D Sharp drop in prices after generic substitution.
June 2006 www.lfn.se (accessed 29 September 2007) [10].

extent be balanced by a switch from patented products to generics, when the price difference increases.[7] However, as the patented products are fully reimbursed there are no strong incentives for doctors or patients to make such changes.[8] It is not possible to give an exact estimate of the net effect, as it is not clear what the market would have been without generic substitution; however, it is safe to assume that the changes would be rather small compared to the observed savings estimated from the price reductions in Table 9.2.

An indication of the importance of the savings is that annual growth in pharmaceutical spending between 1995 and 2005 was only 3.6% in Sweden compared to 4.6% on average in OECD countries. Whilst, in most countries, the cost of pharmaceuticals was growing faster than the total health care spending, the opposite was true for Sweden [11].

So, why was generic substitution a success in Sweden? One important factor for the creation of the image of success was that the patents expired for a number of high-selling products shortly after the introduction of generic substitution. This would have led to significant cost savings, even without the introduction of generic substitution. While generic substitution most likely contributed both to the speed and magnitude of these savings, the incremental effect was smaller than the total estimated savings. In practice, however, generic substitution has been associated with the total change. It is not possible to estimate the incremental effect, as we cannot know what might have happened had generic substitution not been introduced. However, given the high

---

7) A statistical analysis indicates that this effect may have been greater in Sweden 2002–2005 (see Ref. [10]).

8) Recommendations from Drug Therapeutics Committees (DTCs), monitoring of prescription patterns and indicative drugs budgets are used to compensate for the lack of economic incentives for the prescribing physicians.

total sales of those drugs for which the patent expired, there is no doubt that even if the incremental effect had been only half – or even one-quarter – of the total, the additional savings would have been very important.

An additional factor for the success of generic substitution in Sweden is the existence of a public monopoly in retail pharmacy. Savings through a difference between reimbursement and actual price will not be lost as increased profits at the wholesale or pharmacy sector. Apoteket AB, the Swedish drug distribution mo-nopoly, was instructed to stock and distribute the cheapest products, despite the fact that they were only paid for the margin on the actual drugs dispensed. In a free market system, generic substitution will only work if the pharmacist has incentives to distribute a product with a lower price. Thus, the greater the incentive is for generic substitution, the smaller the savings for the payer. The different experience of the effect of generic substitution in Norway illustrates this point. Norway introduced generic substitution at about the same time, but a major part of the savings was lost as increased profits in the distribution system. It is a problem that generic substitution in systems with private distribution systems, generally will lead to smaller cost savings, unless the market is very competitive. There are ways to regulate this, and such regulation was instituted in Norway at a later stage. However, it is important to observe that savings achieved through generic sub-stitutions depend on the incentives present for wholesalers and pharmacists, and a lack of competition in this sector may eradicate any potential gains. This is also seen in evaluations of the effects of the parallel import of pharmaceuticals, where the savings usually remain within the distribution system, and only a minor portion is further transferred to the payer.

A third point – and one which is particularly important with regards to the speed of change – is that price adjustments can be made within a short time. In reference price systems it is administratively burdensome to update prices, and so price changes generally take a long time to filter through. With the introduction of generic substitution, however, the Pharmaceutical Benefits Board (which is responsible for pricing in Sweden) began to review prices once monthly. This speeded up the process of price decline to a rate which was similar to that seen in the US after patent expiration.

It can be concluded that, following its introduction in 2002, generic substitution was an effective policy for achieving both cost containment and improved efficiency in Sweden. The success factors were the timing with patent expiration of major cost-effective drugs with high sales, the institutional factors with Apoteket AB that pass savings to the payer, and an efficient management from the Läkemedelsförmånsnämden (LFN).

9.4
Health Technology Assessment (HTA)

Health technology assessment was developed in the United States during the 1970s as a method to provide information to the Congress for decisions about medical technology. Sweden was one of the first countries in Europe to establish a formal institute for HTA in 1987, first as an agency under the Swedish Government Offices, and from 1992 as an independent public authority.[9] The Swedish Council on Technology Assessment in Health Care, known by its Swedish acronym SBU, was established for ". . . the critical evaluation of methods used to prevent, diagnose and treat health problems" (see www.sbu.se).

While the Ministry of Health may direct the SBU to evaluate specific medical technologies, the aim of the studies is not to be used as a basis for preparing legislation for decisions in Parliament. The reports produced are aimed primarily at influencing clinical and administrative decisions in the health care system. The SBU identifies methods that offer the ". . . greatest benefits and the least risk, focusing on the most efficient ways to allocate healthcare resources" according to the mission statement. The mission does not explicitly state cost containment as a goal, but from the early choices of technologies to evaluate, it is clear that the SBU also sought cost savings from eliminating interventions without any benefits. The first report was published in 1989, examined the use of preoperative routines, and concluded that a number of those could be eliminated for otherwise healthy individuals. A study was also undertaken to evaluate the effects of the report, which showed that annual savings of SEK 235 million were achieved [13]. Another study initiated by the SBU proved that the use of immediate computed tomography instead of admission for observation in patients with mild head injury could reduce costs by one-third, without any negative consequences for medical outcome. A number of other studies have also shown that savings could be made without any expected negative consequences for patients.

However, if we look at the overall number of studies published, and the hundreds of technologies that have been assessed, it is clear that the overall impact in terms of cost-savings is limited.[10] There are several reasons for this. First, in many cases the conclusion of a report is that the evaluated method is underused, and that it is cost-effective to extend the use to relevant patient groups. Second, for many methods – particularly those who have been recently introduced – there is not sufficient evidence to make a clear recommendation about their use, or not. Unfortunately, this is the case for many expensive new drugs, which increasingly have been the focus of HTA studies. During the first 10 years of the SBU, only one major project on drugs was undertaken – an internationally famous study on the treatment of hypertension. Drugs were not a priority area, since a formal assessment process for assessing the efficacy of drugs was already in place for pharmaceuticals through the Medical Products Agency. Hence, the hypertension study focused more on intervention criteria than on the choice of drug for treatment. The study may well have been started

---

9) The first institution established for HTA in Sweden was the Center for Medical Technology Assessment, a joint venture between the county of Östergötland and Linköping university (www.cmt.liu.se). For a more detailed description of the early attempts to HTA in Sweden, see Ref. [12].

10) The not said as an objection to the conclusion that the savings probably far exceed what the government have spent to finance SBU over the years.

with the objective of reducing the number of patients treated for borderline hypertension, but the result of the evaluation did not support that conclusion. Follow-up studies of the hypertension report have focused more on the value of newer versus older drugs, but have failed to provide any conclusions that could have any substantial impact on costs. Rather, patent expiration and price reductions have had stronger impacts on the total cost and cost-effectiveness of treatment.

A third reason that the observed impact on cost containment and efficiency in the health services has been limited, is that the health services lack automatic incentives to use the information provided by SBU. There is, for example, no direct link between the funding of services and the conclusions about their effectiveness and cost-effectiveness in the SBU report. Hence, the implementation must be managed through political, administrative and clinical decision making. Implementation is based on the responsiveness of these processes to act in order to improve quality, effectiveness and cost-effectiveness. It is therefore a time-consuming and costly process to have the evidence implemented in the health care system, where there are no explicit incentives to use available opportunities to contain costs and improve cost-effectiveness. Political, managerial or clinical decision makers are to a very small part accountable for the extent to which evidence-based medicine (EBM) or cost-effectiveness are applied. This does not mean that there is no effect – the effect is simply much more limited than it could have been. This finding cannot be blamed on the SBU, which devotes an increasing share of its resources to market the evidence produced. With a lack of specific incentives, the SBU is reduced to only one of several stakeholders producing information to affect clinical practice and health care management. Even if the authority of the SBU is indisputable, and the reports are very much appreciated, the competition for influence is fierce from many other stakeholders with economic and other interests in resource allocation in the health services; these include not only the suppliers of goods and services but also the doctors, other health care staff, politicians and patients.

9.4.1
Success or Failure?

It is easy to find criteria according to which the SBU has been a great success. First, the high scientific quality of the studies undertaken, and related to this the confidence that the different actors in the health care system have in the results produced. Even the medical professions, who often are critical of initiatives that may influence their autonomy, hold the SBU in high regard. Second, the international standing of the SBU indicates that, in comparison with similar agencies in other countries (who often have considerably more resources), the SBU is in a leading position. Hence, a trust capital has been created during the past 20 years that should be seen as an asset for the future.

It is also possible to find other criteria, however, according to which the SBU may have failed. One criterion is the impact on priority setting and resource allocation.[11]

---

11) This problem is not specific to Sweden (see Ref. [14]).

One of the first studies, the results of which were published in 1989, investigated extracorporeal shock-wave lithotripsy for the treatment of kidney stones. There is no evidence that the SBU report, nor any other studies of the same type, had any significant influence on the spread of this technology. Most County Councils invested in this technology, despite the fact that coordination between them could have saved money and improved outcome. The increased use of percutaneous coronary intervention and stents for the treatment of heart disease is another example. However, perhaps the most significant examples are the introduction of costly new drugs, such as proton-pump inhibitors for gastric ulcers, statins for heart disease, and new cancer drugs such as herceptin. The most striking example is that of tumor necrosis factor inhibitors which, in 2007, were the three most-heavily sold drugs in Sweden. Whilst such widespread use is probably motivated, and takes into account the evidence on efficacy and cost-effectiveness, it does not reduce the need for guidance in how to optimize the use of these drugs in different indications.

It is not that the SBU has not tried to do this,[12] but it is clear that there are two problems that the SBU has difficulty in overcoming. The first problem is to make evaluations in timely fashion (early, before there is much evidence) and to perform them in a very short time. Typically, most SBU studies take several years to complete. The establishment of SBU Alert in 1997 was a response to this problem, although the initial impact seems to have been limited.[13] The second problem is the lack of any systematic approach in the health care system to introduce and evaluate new technologies. Whilst the SBU cannot be blamed for this, its role is limited in a decisive policy field; the introduction, access and incentives for innovation in health care. This is also an issue that is important for both cost containment and efficiency.

9.4.2
How Can the Situation be Improved?

There are economies of scale, both nationally and internationally, to collaboration between agencies that are using HTA studies for policy. In Sweden, such studies are undertaken not only by the SBU but also by The National Board of Health and Welfare, as a basis for national treatment guidelines, and by LFN, the Pharmaceutical benefit Board, for decisions about the reimbursement of drugs. This cooperation provides opportunities for the division of labor, and also means that the studies are used directly for decision making and guidance. For drugs, the Medical Products Agency (MPA) assesses efficacy and safety, both of which are important domains of a core HTA study. Whilst better coordination would mean a more efficient use of scarce resources, perhaps the largest gain in efficiency can be made through international cooperation. Most of the data and studies assessed are international in nature, and are

---

12) For a review of actions taken and examples of impact, see Ref. [12].

13) For an exception, see Ref. [15], which concludes that early warning systems may reduced the diffusion of the technologies studied, while an HTA study in general seemed to have the opposite effect.

also assessed by similar agencies in other countries. A division of labor in putting together evidence that is not context-specific could save much duplication, not only for governments but also for the producers of technologies who need to documents their effects.

A more efficient accumulation and assessment of data that is not context-specific would give more room for developing data that are of specific importance for defined decision makers. Much of this context-specific data relates to the economic consequences of new technologies. These in turn depend on the health care system and the practice patterns used within these systems. It is therefore logical that the SBU in the new strategy for 2007–2010 proposes greater resources to health economic studies.

One problem with such studies is that it is seldom possible to use the traditional HTA method of literature searching and synthesis of information, as often there are no relevant studies to find and synthesize. Thus, original data must be collected, and it is therefore important that access to relevant data in the health care system is improved. Of particular importance here is the successive accumulation of evidence in the health care system from the use of new technologies. Such studies have been successfully set up for new drug treatments in diseases such as multiple sclerosis (MS) and rheumatoid arthritis (RA). It is also rather common that reimbursement decisions made by LFN are linked to the collection of further data to establish the use, effectiveness and cost in clinical practice. These data may then be used for a revised decision at a later time, usually after two years.[14]

In order to be more useful for policy and management decisions in the health care sector, studies need to be timely, based on relevant data, and conducted with competence to ensure high quality. However, this is not generally enough, and the link between the study results and resource allocation in health care must also be strengthened. This is probably the most difficult part, as it involves not only making study results available and understood but also including the use of appropriate incentives and management processes to make sure that the potential for improvements in efficiency and cost containment is materialized.

Management of the cost and cost-effectiveness of new technology through the accumulation and publication of evidence is a complicated process, and HTA is not a perfect instrument. But, is there a better alternative? Significant progress has been made to create studies that would have a better impact on the allocation of resources. However, the 'Achilles' heel' here seems to be implementation in the resource allocation processes in the County Councils, and particularly in the hospitals. There are options to improve this situation, nonetheless. While hospital inpatient days continue to decline, the hospital is the key to the introduction of new technologies,

14) It may be argued that the establishment of LFN is the single most important step in Sweden to make HTA studies policy relevant, and thus have a direct impact on open priorities and resource allocation; similar to the importance of establishment of NICE in UK in 1999. While LFN is not a traditional HTA agency, they use HTA methods, including cost-effectiveness studies for both individual reimbursement decisions and review of whole drug classes. LFN has obviously benefited from the previous investments in HTA and studies and earlier as well as present studies carried out by SBU.

and increasingly also for new drugs. This will provide future opportunities to coordinate the introduction and also undertake follow-up studies on the effectiveness and costs in clinical practice. Sweden has unique opportunities to follow patients through the health care system to document both costs and outcome. With more resources spent on chronic illnesses such as cancer, RA and brain diseases (MS, dementia, Parkinson's disease, depression, etc.) it is increasingly important to undertake long-term follow-up studies as a basis for HTA studies. The wider use of local and national data may also be a key factor to improve the impact of HTA studies for management and clinical decisions.

9.5
The Current Situation

In 1980 Sweden spent 8.3% of GDP on health care, and 8.4% in the year 2000. By the year 2005 (when the latest OECD data were available), the share was 9.1%, which is more or less exactly what could be expected given the level of GDP per capita in Sweden. Whilst it is difficult therefore to imagine that there would be any major problems with cost containment in Sweden, behind these favorable macro-statistics there are several factors that must be observed. Although, at present, the County Councils are in a good financial situation, they have for the major part of the past decade been running in deficit, and there are no reserves for the future. With a downturn in the economy, and the consequent reduced employment and incomes, the tax base will be directly affected and the need to contain costs will become immediate unless Central Government steps in. Although the public finances are strong at present, an economic downturn will also put pressure on Central Government since, in such a situation, health care must compete with many other areas of public expenditures. It is necessary therefore that the County Councils develop a clear strategy for such a situation, even if it does not appear to be imminent.

The more pressing problem is the need to improve efficiency and quality in the health care system. Most of these investigations are conducted at the micro level in the hospitals, where the major portion of the health care resources is spent. The improvement of processes in order to reduce costs and improve quality has developed as an increasingly important task for management. A shift in focus from incentives and the individual behavior of physicians with regards to the use of resources to the management of processes within the hospital can be observed. However, whilst it remains an open question as to how successful this policy will be, efficiency and quality are clear issues where management will have a decisive impact.

The further development of the DRG classification system and the cost-per-case database is of strategic importance for several reasons. First, it is important to obtain an accurate description of the activities in inpatient and outpatient care, in order to allow comparisons of productivity between units and over time. Second, it is important for the further development of a system of fair competition between providers – both public and private – for the resources available for public health care.

Linked to this is the goal of an improved choice of provider for the consumer/patient. An understandable, verifiable and effective reimbursement system is vital to the success of these goals.

While it is easy to admire the systematic development of the DRG and KPP projects in Sweden, the problem is that such systems take a very long time to develop, and their implementation is uncertain. Yet, 'the best is often the enemy of the good', and the Ministry of Health has now taken the initiative to speed up the process (although experience tells that it is unlikely to be successful in the near future). A reimbursement system cannot be fully developed before being implemented; rather, implementation must come together with strong management, to make the necessary adjustments. These adjustments are continuously needed in order to develop a functioning system that adjusts to the rapid development in health care. A rigid application of DRG reimbursement, without adjustments to technological change, can have a negative impact on innovation and dynamic efficiency. So far, we have not identified any major problems with this approach, as indicated by the speed in which both laparoscopic surgery and drug-eluting stents have been introduced, despite significant cost increases for the hospital, with savings and quality improvements mainly occurring after the patient has been discharged from hospital.

Although, until now, generic substitution has been a great success, its importance will be lessened in the future, for several reasons. One reason is the fact that fewer drugs with large sales in ambulatory care will lose their patent in the future. Several of the drugs for which the patent expires are also hospital products (e.g. some anticancer drugs), where the successful model with substitution at the pharmacy is not applicable. In addition, many of those drugs with high sales that will lose patent are known as 'biosimilars' – that is, large molecules that cannot be copied in exactly the same way as small molecules. These products are also mainly used in hospitals or for outpatients. When looking to the future, new drug introductions with high sales have changed from the ambulatory to the hospital setting, indicating that it will be in the hospital that the decisions about cost containment and cost-effectiveness will increasingly be made.

A specific Swedish perspective is also the institutional changes in retail pharmacy which are expected. The monopoly of Apoteket AB is under review, and different models for the re-regulation of drugs have been put forward. Even if there is a great awareness about the benefits of keeping the present system for passing savings from generic substitution back to the payer, any changes towards creating a competitive retail market will mean that the payer will have to share these saving with the pharmacies.

At present, the single most important issue related to efficiency and quality, as well as cost containment, is the introduction of new (often expensive) medical technology. Although this is mainly a management problem, there are also wider implications for health policy. Ultimately, it will be a question of how much a nation can afford and choose to spend on health care, and about incentives for investments in medical research to develop new treatment opportunities. Moreover, it is an issue with national as well as international implications, and there is also a link to the development of the biomedical industry as a source of economic growth in a country.

This issue also relates to the fundamental objective for the health care system, namely that everybody should have the same right to treatment, regardless of income and wealth. While reimbursement is therefore a key factor for access, the new technologies are often so expensive that even third-party payers, both public and private, face difficult decisions about what to pay for.

Health Technology Assessment, reference pricing and generic substitution, and DRG-based reimbursement of hospital care are the three policies which serve as focal points in helping the policy makers and managers in the health care sector to address issues related to the introduction of new technologies. HTA will most likely become the most important policy instrument, as it can help in the 'great trade-off' between static and dynamic efficiency, as well as provide optimal incentives for innovation while controlling costs. However, this approach requires that the politicians governing the County Councils fully embrace the principles of open priority setting, and that managers and leading clinicians at the hospitals take leading roles in their implementation. Reference pricing and generic substitution are aimed at containing the costs and improving the cost-effectiveness of off-patent drugs and other technologies, thus providing headroom for investments in new innovative technologies. DRG-based reimbursement may also be the key to the development of 'pay-for-performance' systems, which represent another major objective for health care payers in the future.

References

1 Saltman, R. and von Otter, C. (1987) Revitalizing public health care systems: a proposal for public competition in Sweden. *Health Policy*, **7**, 21–40.

2 Arvidsson, G. and Jönsson, B. (1997) Politik och marknad i framtidens sjukvård, SNS Förlag.

3 Culyer, A.J., Evans, R.G., Graf von der Schulenburg, J.-M., van de Ven, W. and Weisbrod, A.B. (1991) International review of the Swedish health care system. Occasional paper no. 34. SNS, Stockholm.

4 Hälso- och sjukvården i framtiden – Tre modeller (1993). SOU 1993:38. (Swedish Health Care in the Future: Three Models).

5 Håkansson, S. (2000) Productivity changes after introduction of prospective hospital payments in Sweden. *CASEMIX*, **2** (2), 47–57.

6 Håkansson, S., Paulson, E. and Kogeus, K. (1988) Prospects for using DRGs in Swedish hospitals. *Health Policy*, **9**, 177–192.

7 Gerdtham, U., Rehnberg, C. and Tambour, M. (1998) Estimating the effect of internal markets on performance in Swedish health care. *Applied Economics*, **31**, 935–945.

8 Lundin, D. (2000) Moral hazard in physician prescription behavior. *Journal of Health Economics*, **19**, 639–662.

9 Ekelund, M. (2001) Generic entry before and after reference prices, in *Competition and Innovation in the Pharmaceutical Industry* (ed. M. Ekelund), EFI, The Economic Research Institute at the Stockholm School of Economics (www.hhs.se/efi) (Dissertation).

10 Engström, A., Jacob, J. and Lundin, D. (2006) Sharp drop in prices after generic substitution. www.lfn.se (accessed 29 September 2007).

11 OECD health statistics (2007) (www.oecd.org).

12 Carlsson, P. (2004) Health technology assessment and priority setting for health

policy in Sweden. *International Journal of Technology Assessment in Health Care*, **20**, 44–54.

13 Brorsson, B. and Arvidsson, S. (1997) The effect of dissemination of recommendations of use. Preoperative routines in Sweden, 1989–91. *International Journal of Technology Assessment in Health Care*, **13**, 547–552

14 Olivier, A., Mossialos, E. and Robinson, R. (2004) Health technology assessment and its influence on health care priority

setting. *International Journal of Technology Assessment in Health Care*, **20**, 1–10.

15 Packer, C., Simpson, S., and Stevens, A. on behalf of EuroScan: the European Information Network on New and Changing Health Technologies (2006) International diffusion of new health technologies: a ten country study of six health technologies. *International Journal of Technology Assessment in Health Care*, **22**, 419–428.

Index

Cost Containment and Efficiency in National Health Systems: A Global Comparison
Edited by John Rapoport, Philip Jacobs, and Egon Jonsson
Copyright © 2009 WILEY-VCH Verlag GmbH & Co. KGaA, Weinheim
ISBN: 978-3-527-32110-0